DATE DUE			
JA 25 '98			

OPPORTUNITIES IN CIVIL ENGINEERING CAREERS

D. Joseph Hagerty
Louis F. Cohn

Forword by
Dan Barge
President
American Society of Civil Engineers

VGM Career Horizons
A Division of National Textbook Company
4255 West Touhy Avenue
Lincolnwood, Illinois 60646-1975 U.S.A.

Cover Photo Credits

Front cover: upper left, American Society of
Civil Engineers photo; upper right, NTC
photo; lower left, American Society of Civil
Engineers photo; lower right, Army Corps of
Engineers photo.

Back cover: upper left, Los Angeles
Department of Water and Power; upper right,
DuPont, Inc.; lower left, British Columbia
Hydro, Vancouver; lower right, American
Society of Civil Engineers photo.

ABOUT THE AUTHORS

D. Joseph Hagerty is professor of civil engineering at the University of Louisville. He has written widely, publishing seven books and over 65 journal articles on various facets of civil engineering, and he is a consultant to governments and industry throughout the United States.

Louis F. Cohn is professor and chairman of the department of civil engineering at the University of Louisville. He has published extensively and lectured on transportation engineering issues all over the world. Dr. Cohn has performed sponsored research and consulted for many state and federal agencies.

ACKNOWLEDGMENTS

The authors wish to acknowledge the careful and patient assistance of Ms. Judy Adams and Ms. Laura Helm who typed and corrected the manuscript of this volume. Mrs. Michael Greenwell, Ms. Brenda Hart, and Ms. LeAnne Whitney performed similar services for the first edition of this book.

For information on their programs, the following individuals are gratefully acknowledged:

Dr. Frank K. Bogner, University of Dayton

Ms. Doreen Ross, JETS

Dr. Paul Zia, North Carolina State University

FOREWORD

Choosing one's profession and career path is a momentous decision. For some, it is a natural evolution, such as the fortunate few who grow into the family profession and thereby have some unique opportunities. But for most of us, it is a deliberate choice, seldom made earlier than high school age. More likely, it will occur midway in the undergraduate college period. By that time, one is likely to be greatly influenced by his or her peers, professors, or role models.

Drs. Hagerty and Cohn's treatise is an excellent tool for evaluating those elements upon which a prudent decision should be made. It is factual, thorough, and in sufficient detail that a close study will provide the reader a good knowledge of the many opportunities for a productive career in civil engineering. It also outlines the further opportunities for advancement through graduate studies, continuing education, and association with professional and technical societies and associations. All of these activities are most helpful in one's professional development.

From the viewpoint of a civil engineer whose career spans more than four decades, the reward of knowing that projects to which one has contributed have been and will continue to be of service to the public is more important than the compensation received. Civil engineering is indeed a people-serving profession.

Dan Barge, Jr.
President
American Society of Civil Engineers

v

PREFACE

This book is about civil engineering and the opportunities for a career in this field. The book is intended to help you learn about engineering and technology in order to help you make choices for a future career. We have emphasized the requirements for such a career—education, aptitude, and hard work—and the returns you will gain if you invest time and effort in this field—reliable employment at a good salary, professional recognition, and the chance to improve the quality of life for yourself and others. Before we begin to talk about this subject in detail, we must define some basic terms.

One of the most frequently used words in today's society is technology. To many people, technology is a monster responsible for most of the evils present in the world today. To others, technology is a miraculous means for obtaining all they want out of life. Just what is *technology*? Very simply stated, technology is the application of scientific knowledge to solve problems. Most of the people working in technological fields today would state that they are trying to apply scientific knowledge to create a better life for their fellow human beings, to use natural and human resources wisely.

Applications of scientific knowledge to solve everyday problems create a need for more applications of scientific knowledge. In other words, technology feeds on itself. That is not to say that technology needs to be more complex or more sophisticated. Rather, technology must always improve. Improving technology will require many new engineers and scientists. The increasing influence of technology on our lives will also require tremendous efforts from non-scientists to guarantee that technology always improves the quality of life. In order to improve, technology must always be made more humane.

What will your role be in tomorrow's technology? You will be involved in technology, whether you want to be or not. It is important for you to think now about what you want to do in the coming years. Not everyone is suited to a career as a scientist or engineer. How can you judge your own potential for a career in engineering, and in particular in civil engineering? What factors determine success or failure in the civil engineering profession? An individual must possess certain inherent abilities in order to be successful as an engineer or scientist. Also, it will be necessary to devote considerable time, effort, and attention to your profession in order to be a successful civil engineer.

No author could hope to describe all the applications of scientific knowledge and practical experience involved in modern technology. It is impossible to anticipate the changes in technology which will occur in coming years. Between the time this book was written and the time you are reading these words, changes occurred which were impossible to predict. We are not trying to tell you everything you could ever learn about technology, about engineering, or about civil engineering. We hope merely to contribute in one particular way: by informing people who are seriously investigating careers in civil engineering. We hope to help you make a wise career choice. In the following pages, we will try to tell you what civil engineering is and what civil engineers do. We hope to give you enough information so that you can decide whether or not civil engineering is the career for you. We hope that some of you will join us and the other members of our profession in our efforts to understand our world better and to improve the quality of life for its inhabitants.

Dedicated
to
John E. Heer, Jr., Teacher and Friend

CONTENTS

   Engineering. The scientist. The engineer. The
   technician. The craftworker. Perspective. Some civil
   engineering history. Future outlook. Employment.
   Salaries. Minorities and women. Contributions—now
   and in the future. Obligations. The impact of
   computers. Artificial intelligence and expert systems.
   Personal attributes.

   High school. Choosing a college or university.
   Curricula. Bachelor's degree programs. Options within
   basic programs. Advanced study in engineering.
   Cooperative work programs. Transferring to other
   professional schools. Graduate work in other
   disciplines. Continuing education in engineering.
   Education costs. Financial aid opportunities.

A civil engineer inspects plans during the construction of the Intermountain Power Project in Utah. (Los Angeles Dept. of Water & Power photo)

CHAPTER 1

THE SCOPE OF CIVIL ENGINEERING

Engineers are persons who utilize natural resources to improve human living conditions, as mentioned very briefly in the introduction. To make decisions about your career, you will need a more detailed and comprehensive definition of engineering and, more specifically, civil engineering. Engineering is closely related to the broader field of technology, which is the application of science to the solution of real problems. Since technology involves the application of science and scientific principles, it is obvious that scientists also are involved in technology. Their work must relate closely to the work of engineers. The important factor here is that technology is *applied* science. The application of scientific principles through technology includes the use and handling of natural materials to create a variety of things to satisfy human needs. Quite often this creation of material goods is the job of the craftsperson. In many instances, the craftsperson is directed in this work by the engineering technician. Somewhere between the scientist searching for absolute truth and the craftsperson dealing with natural materials such as wood, concrete, and steel, the engineer functions as a key member of the technological team. What are the characteristics which distinguish the engineer from the scientists, technicians, and craftspersons with whom he or she works? First, let's take a closer look at engineering.

ENGINEERING

Perhaps the best way to answer the question "What is engineering?" is to look at the answers to this question which have been given collectively by practicing engineers. The American Society for Engineering Education has given the following definition of engineering:

> Engineering is the profession in which a knowledge of the mathematical and natural sciences gained by study, experience, and practice is applied with judgement to develop ways to utilize the materials and forces of nature, economically for the benefit of mankind.

Obviously, the engineer is deeply involved with the application of mathematical and natural sciences. Because of this, many people are confused about the distinction between scientists and engineers. The confusion increases because many individuals function as both engineers and scientists. They apparently perform engineering work, applying scientific principles to the solution of practical problems, but they do not even call themselves engineers. They do not possess the general characteristics of a professional engineer: a college degree, an engineering license or a legal registration, and membership in a professional engineering society. Other individuals who were originally trained in engineering enter other fields of activity but continue to consider themselves engineers. Many people who function as sales representatives or as managers refer to themselves as "engineers" because their basic education and early experience were in engineering. Finally, many people, especially children, are confused by the application of the term "engineer" to persons who drive trains or who are otherwise engaged in activities far removed from professional engineering. Inevitably, some individuals will function as scientists, engineers, and technicians all at the same time. Others will sometimes act as engineers and at other times as technicians. For the moment, let's disregard these people who are sufficiently flexible to practice in several different fields, and concentrate on the characteristics of engineering mentioned above.

First, engineering is a profession. When we say that, we mean a vocation or a calling which requires the application of specialized knowledge gained through long and intense preparation. For most people, this preparation for a professional career is obtained through formal education. Two clear examples of professions other than engineering are medicine and law, which require long years of educational preparation and specialized training for successful practice.

In addition to intensive and specialized preparation through education, a profession has other major characteristics. One of these characteristics is the insistence on high standards of conduct by the members of the profession themselves. Members of the profession form some sort of organization to police themselves and their colleagues and insist upon high standards of ethics, as well as achievement in technical subjects, by their fellow professionals. Maintaining high standards of technical achievement often requires members of a profession to continue to study and gain experience in their chosen fields throughout their entire professional careers.

Finally, most people who consider themselves professionals would describe themselves as engaged in some form of public service. In other words, the underlying purpose of most professions is collective action by the members of that profession for the public good. It is easy to see that the day-to-day activities of doctors and lawyers are directed toward improving human living conditions and developing a more civilized society. In much the same way, engineers work to develop a better quality of life for their fellow human beings. To do this, engineers apply scientific principles and their experiences in using natural resources and the forces of nature for the benefit of humankind. Many engineers are employees of large corporations or consulting firms and, as such, their contributions to the public good may be somewhat indirect and therefore hard to recognize. Nevertheless, the effects of their activities upon community well-being are quite important. In any of the industrialized nations of Western Europe and North America, almost everything that the average citizen uses, touches, or eats during her or his daily life has been influenced in a significant way by an engineer's activities at some stage in its manufacture or processing.

THE SCIENTIST

When the members of a football team gather in the locker room just before the beginning of the game, they all have one common goal: to take the football across the opposite goal line and to keep their opponents from scoring. However, the individual roles of the members of the team are quite different in accomplishing this common objective. A defensive tackle concentrates on identifying and tackling the ball carrier from the other team and does not think about catching passes unless he is lucky enough to intercept a poorly thrown pass from the opposing quarterback. The quarterback's head is filled with a great number of offensive plays in which he will hand the ball off to another back, throw a pass, or carry the ball himself. The quarterback is not thinking about tackling anybody, and it is usually a disaster on the field when the quarterback ends up tackling a ball carrier from the other team. In much the same way that the individual members of the football team all have different roles to play in accomplishing a common goal, the various members of the technological team have individual specialized roles to accomplish. The scientist can be imagined sitting on the far end of the bench from the technician or the craftsperson. The scientist's basic purpose is to discover knowledge, to uncover new facts, and to develop more comprehensive understanding about the natural world. The scientist also attempts to explain the facts that he or she discovers by developing theories. These theories relate causes and effects in the natural system the scientist has observed and investigated. These theories allow the scientist to predict the results of carrying out a particular sequence of actions. The scientist seeks to *know* rather than to *apply,* in most instances. Of course, many scientists attempt to apply their discoveries for beneficial purposes. In this way, their activities overlap with those of engineers. However, the scientist's principal concern usually is not the application of new knowledge, but simply the discovery of that knowledge itself. Sometimes scientists are called "applied scientists" because they are interested in developing applications of scientific methods, but the basic activity of even these scientists remains the discovery of new knowledge.

THE ENGINEER

Most of the time, the engineer is not interested in the discovery of new knowledge through scientific investigations. The engineer is more closely concerned with the application of existing knowledge. Sometimes, it is true, an engineer will become involved in scientific investigations in order to solve particular problems. Most of the time, however, the engineer will design or plan systems to achieve certain definite results. These systems may include production processes, operation and maintenance procedures, etc. In all of her or his activities, the engineer is trying to achieve results that will benefit society. However, simply saying that the engineer wants to benefit society does not give a full description of the way most engineers work. The engineer tries to achieve benefits for the public at a minimum cost in money, materials, and time. As the old saying goes, "An engineer can do in an hour for ten dollars what any fool can do for a fortune during a lifetime." An engineer goes to great trouble to forecast the behavior of the systems he or she has designed in order to achieve the most efficient results from the investment of materials and effort in the system. The engineer always tries to evaluate quantitatively the benefits and costs of the activities included in any process he or she has developed. In general, the engineer does not assume immediate responsibility for seeing that her or his designs and plans are carried out. This responsibility is the job of the technician.

THE TECHNICIAN

Whereas the engineer is concerned chiefly with *conceiving* and *designing* plans and systems, the technician is concerned principally with the actual *doing* and *making* involved in the design. For example, the technician may be responsible for making time-and-motion studies in an effort to develop the most efficient process for carrying out the design developed by an engineer. A technician may be the supervisor on a construction project where the plans for the building have been drawn up by an engineer and an architect. The technician will be responsible for seeing that the construction materials are assembled in the way that the engineer has specified in the contract documents. In general, the technician is more concerned with the hands-on application of experience or scientific principles to achieve a particular result than is the engineer, who devotes more attention to the overall planning of the entire system. The technician is responsible for seeing that the designs and plans developed by the engineers are implemented, often with machines and materials, by craftworkers.

THE CRAFTWORKER

The skilled craftworker is at the opposite end of the technology team bench from the scientist. The craftsperson uses her or his hands and skills just as much as he or she utilizes mental processes. He or she is seldom called upon to utilize scientific principles or scientific knowledge. He or she is much more likely to use tools with a skill developed through experience gained during a lifetime on the job, or through tips from co-workers. He or she must have simple manual dexterity and an inner feeling for material behavior and performance. Craftworkers include electricians, machinists, carpenters, masons, and many others. Craftspersons often talk about having a "knack" for their particular jobs. This refers to the element of skill and talent included in their work. The importance of the craftsperson as a member of the technology team should not be underrated. To a great degree, the overall success of many technical systems depends on the skill of the craftsperson in using tools to build and maintain engineered systems. The designs of the engineer will not be successful if the craftsperson is not concerned and careful in the construction of the designed facility. There is no substitute for proficiency in construction and in proper use of completed systems. The work of the craftsperson is extremely important to the functioning of those systems.

PERSPECTIVE

The activities of the engineer can now be seen as part of the team of specialists engaged in applying scientific knowledge and practical experience to the solution of technical problems. The role of the engineer also must include considering many non-technical factors in developing plans and designs. These non-technical factors include legal restraints on proposed activities as well as the impacts on society which would be produced if proposed designs and plans were actually carried out. The engineer must also consider economic factors and try to develop the most efficient and least expensive answer to a particular problem. Finally, the engineer must be aware of aesthetic considerations. The best designs and plans will not only be useful but will also reflect a proper appreciation for style and grace. The best use of natural forces and materials will be attained through structures, machines, and facilities which are beautiful as well as functional.

SOME CIVIL ENGINEERING HISTORY

If we apply the definitions of engineering given above and concentrate on the applications of experience to the solution of practical problems, many of the achievements of ancient peoples could be considered engineering works. The Stone Age people of Switzerland who drove timber logs vertically into the bottoms of lakes in order to move their villages away from the land and out of reach of marauding tribes certainly could be considered engineers, by virtue of their clever construction technique and the use to which they put their ancient pile foundations. Certainly the Egyptians with their advanced methods of surveying and land measurement employed engineering techniques. The building of the Great Pyramids still could be considered one of the major engineering achievements of history. The building of Stonehenge, lately discovered to be a sophisticated astronomical observatory, certainly would rank as a major engineering achievement even today. However, in many of these early activities the work of the engineer was impossible to distinguish from that of the technician and the craftsperson. The application of scientific principles in engineering work did not really begin until after the scientific revolution and the birth of scientific inquiry in the fourteenth and fifteenth centuries. Prior to this time, there was a wide gulf between science—theoretical knowledge and analysis—and practical engineering work.

Galileo was among the first of the scholars of the sixteenth and seventeenth centuries who made important scientific discoveries through experiment and investigation. Leonardo da Vinci's investigations included tests on the strength of materials, analysis of pulleys and levers, study of the forces acting on an arch, and many other engineering applications. The work of these men was succeeded by that of many other early scientists including Descartes, Mariotte, Newton, and Euler. However, the link of engineering to science was not made in the same way in all countries. In France, the theoretical basis for engineering was advanced significantly because of the royal encouragement given to science and higher education by the king.

One of the outstanding figures of the seventeenth and eighteenth centuries was King Louis XIV of France. He is important to the growth of engineering and particularly to the applications of science in engineering because he was persuaded by his minister Colbert to establish academies of science and architecture in the late 1600s. Later he established the Corps de Genie, a group of military engineers. The king decreed that this group of engineers was to be given a *scientific* education under the direction of Vauban, the foremost military engineer in France at that time. By establishing this engineering corps for the French army, King Louis set a trend calling for the scientific education of engineers and thus emphasized the application of scientific principles in engineering practice.

During this time and in the next century, while French engineers were being educated in a more scientific fashion, their counterparts in the British Isles were still relying heavily on experience and field training. For example, British engineers during the seventeenth century learned much from the Dutch engineers brought into England by King William to assist in major drainage projects. During the eighteenth century, the development of this group of British engineers devoted to civil works continued, and led to the formation of what can probably be considered the first group of professional engineers. This occurred when an engineering society was formed by John Smeaton in 1771. Smeaton was a pioneer in the field of civil engineering not only because he formed the first society of engineers engaged in civil works, but he apparently also was the first engineer to use the title "Civil Engineer" behind his name to designate his profession. Smeaton was also the first individual to appear as an expert witness to give testimony in a lawsuit, where the verdict was greatly influenced by technical opinion. Formal education in engineering was established at the University College of London in 1828 and later in other schools of the British Isles. In France, the Ecole des Ponts et Chaussées had been founded in 1747 for the express purpose of training engineers. One of the most important features of this school was the emphasis given to physics and mathematics in the education of engineers. This emphasis on scientific education for engineers was continued by Napoleon who established the Ecole Polytechnique in 1795. At this school, future engineers were given a broad education in the sciences, and graduates were prepared for advanced studies at the Ecole des Ponts et Chaussées. These French schools became the models for numerous engineering schools started in Europe during the nineteenth century.

Because of the fundamental difference between the educational backgrounds of British engineers and continental engineers, the young continental engineer, after finishing his formal education, was far more advanced in scientific theory and design than was his British counterpart. In contrast, the continental graduate had little or no experience. The great British engineers of the nineteenth century in general were naive geniuses, with little appreciation for scientific principles. They began as apprentice instrument makers, millwrights, or masons, and were more closely linked to the ancient masters of the Middle Ages than to modern-day scientists. Because of the obvious links between the American colonies and the mother country of Great Britain, many of the early engineers in the American colonies relied more heavily on experience than on scientific theory. However, the growing estrangement between the colonies and

Great Britain, which culminated in the American Revolution, produced an introduction of French scientific principles in engineering education in the United States.

In the infancy of the United States the emphasis in engineering was placed on military engineering. The Continental Congress authorized an engineering corps for Washington's army in 1775. After the Revolution, this emphasis on military engineering changed. In 1794, Congress authorized President Washington to recruit a body of engineers to design and construct fortifications around harbors on the Atlantic seacoast. This work led to the improvement of the harbors themselves, a project of civilian rather than military character. In 1807, Congress recognized the need for a survey of the coastline of the United States and appropriated funds to employ a group of surveyors and engineers for this task. Finally, in 1824, Congress approved funding for surveys for roads and canal routes which were considered important by the President for the young country's growth toward the west. These surveys were to be carried out by the fledgling United States Corps of Engineers. This increasingly important civilian role in engineering continued to be evident with the coming of the railroads in the mid-nineteenth century and during the development of the Industrial Revolution.

Today, civil engineering stands apart on its own, quite distinct from military engineering and from other branches of engineering. Basically, civil engineering consists of the planning, designing, construction, and operation of physical facilities and structures. These facilities may include bridges, highways, airports, and other facilities for transportation. Civil engineers plan, design, and supervise the construction and operation of dams, floodwalls, irrigation works, harbors, and other waterways. Obtaining and purifying water supplies for the general public, as well as treating wastewaters, are included among the principal activities of civil engineers. Civil engineers are active in the design and construction of industrial buildings, power plants, office buildings, and major residential complexes. In all of these areas, civil engineers are also active in research designed to expand the body of knowledge which presently exists. In

approximately 24 percent of the total to only about
These changes in enrollment reflected an emphasis on
ng projects in the national economy during those
he 1980s, in contrast, much more emphasis has been
ications of electronics—in particular, computer
every facet of American and world society. A study
c Manpower Commission (*Manpower Comments,*
9, November 1985) indicated that of all of the engi-
ts working in the United States in 1984, more than
sidered computer specialists. Employment in this
doubled during the ten years from 1975 to 1985.
out 250,000) was substantially greater than the
earning degrees at all degree levels in all computer
the United States (about 100,000). This indicates
ct: many other engineers have become closely as-
uter applications and with other applications of
n addition to electrical and electronics engineers
tists. However, to the general public, the branch
closely associated with electronics and comput-
eering. As a result, many high school students
trical engineering curricula during the 1980s.
hosen not to go into civil engineering or chemi-
gh the mistaken idea that engineers in those
omputers and are not intimately involved in
and development. In 1984, civil engineers ac-
ut nine percent of the total undergraduate
ng schools in the United States. Chemical en-
r only about seven percent of the total
ent. The Scientific Manpower Commission
the same volume of *Manpower Comments*
ers from U.S., state, and local governments
sed 72 percent in 1985 as a result of the need
s such as highways, water supply systems,
t facilities. They reported that the United
otection Agency predicted unprecedented
solve environmental problems, including
ater treatment, developing methods for

this research, they work with such scientists as geologists and physicists, non-technical personnel such as economists and city planners, and engineers from other branches of the profession.

FUTURE OUTLOOK

We've talked about the history of engineering, and the growth of civil engineering. What does the future hold for civil engineers? Will they continue to play an important role in coming years? What are the prospects for steady employment, for good salaries, and for important responsibilities in the future?

EMPLOYMENT

The *Occupational Outlook Handbook* published by the United States Department of Labor, Bureau of Labor Statistics, contains the following statement in the 1984-85 edition:

> Employment of civil engineers is expected to increase faster than the average for all occupations through the mid-1990's

Studies by the Department of Labor indicated that employment opportunities for all engineers will be good through 1995. Most replacement openings will be created by engineers who transfer to management, sales, or other professional occupations, rather than by those who retire or die. Throughout the 1980s and 1990s engineering will continue to be a popular profession. At present in the United States, engineering is the second largest profession; engineers are outnumbered only by elementary and secondary school teachers, of whom there are almost 3,000,000 in the United States. Previous studies by the Department of Labor indicated that the fastest growing occupations through the 1970s and 1980s would continue to be professional and technical, the ones requiring the most extensive educational preparation. The number of engineers in the United States increased from approximately 870,000 in 1960 to more than 1,000,000 in 1970 and to almost 1,200,000 in 1982. Of this number,

600,000 engineers are employed in manufacturing operations or associated activities, 400,000 in non-manufacturing activities, and 160,000 are employed by local, state, or federal governments. More than 400,000 engineers are legally registered in the United States. (For more on registration see Chapter 6.) The total employment of engineers in the United States in the early 1980s, according to the United States Department of Labor, was divided in the following way:

Electrical Engineers	320,000
Mechanical Engineers	210,000
Industrial Engineers	155,000
Civil Engineers	155,000
Chemical Engineers	55,000
Aeronautical Engineers	45,000
Petroleum Engineers	25,000
All Other Engineers	225,000

Of the 155,000 civil engineers counted during the 1982 census by the United States Department of Labor, approximately 40 percent were employed by governmental agencies, about one-third in private consulting firms, and the remainder in construction activities, by public utilities, by railroads, and by manufacturing concerns.

A 1983 study by the National Science Foundation reported in the February 23, 1983, issue of *Science Resources Studies Highlights* included predictions of growth in technical employment as well as economic growth for the country as a whole. This study produced two estimates of growth: a low estimate based upon a growth rate of about 1.6 percent each year in the Gross National Product (a good indicator of economic health), and a high projection of growth of 4.3 percent per year in Gross National Product. Using these ranges in predicted growth for the economy of the United States, the study authors predicted a growth in employment of scientists, engineers, and technicians varying from 2.5 percent to 4 percent per year through the 1980s. The predicted total employment in 1987 of scientists ranged from 830,000 to 880,000. Predicted 1987 employment figures for engineers ranged from 1,300,000 to 1,420,000. The

predicted growth in employme
total employment of between 1
particular, the increased civil e
ed by the study authors to ran
180,000 by 1987.

There is always a time lag
jobs available in any field a
to study in those fields wl
with the opening up of spa
ed States was falling behi
great growth in engineeri
corresponding increase
programs during the 1
study declined slightly
its lowest level in 20 ye
orous growth in eng
increase in enrollme
crease in enrollment
seen a decrease in e
engineering, with a
engineering. Elec
86,146 full-time
1984. During the
rollments decre
Mechanical eng
69,608 in 198
ments droppe
enrollments c
What is the r

There are
occupations
engineering
1970s as a
vironmen
carried o
cent of t
students

hazardous waste control, developing systems for groundwater management, and constructing facilities for municipal waste disposal.

The net result of this decrease in enrollments in civil engineering may be unfortunate for the country as a whole, but it will be very advantageous for those students deciding to go into civil engineering. Projections by the United States Department of Labor indicate an increase in civil engineering jobs of at least 1.5 percent per year. With increases in available jobs, and decreases in the number of students enrolled in civil engineering, the demands for graduates will be high and the supply will be low. Students graduating from civil engineering curricula in the late 1980s and early 1990s will be faced with a pleasant prospect of multiple job offers at attractive salaries.

SALARIES

Salaries and other rewards will be described in more detail in Chapter 4, but at this point we can say a brief word about salaries in civil engineering. Starting salaries for engineers with bachelor's degrees have risen steadily from an average of about $3,000 per year in 1945 to more than $13,000 per year in 1975 and to more than $27,000 per year (average starting salary) in 1985. Of course, much of this increase in salaries can be attributed to the increase in cost of living. However, even if the increases in the cost of living over the last 30 years are taken into account, it is apparent that beginning salaries for engineering graduates today are much higher than the salaries paid to new engineering graduates in 1945. During the last ten years, the real buying power represented by engineering salaries has remained relatively constant. From 1972 to 1984, for example, the Consumer Price Index, an indicator which accounts for increases in the cost of living, increased by an average annual rate of 7.9 percent. In the same period, starting salaries for engineers increased at an average rate of 7.7 percent each year. This indicates that engineers' buying power did not decline in those times of high inflation; unfortunately, this statement cannot be made about the salaries of many other occupations in the United States during the same time period. The Consumer Price Index is based upon 1967 as a base year; in other words, prices in recent years are given as a multiple of the prices in 1967. For example, in 1972, it would have cost you $1.24 to buy the same goods or services that you could have purchased for $1.00 in 1967. In 1980, the same goods and services would have cost you $2.36. In 1984, the goods or services that you could have purchased for $1.00 in 1967 would have cost $3.07. The average starting salary of engineers in 1984 was about $26,100, but in terms of 1967 dollars this same salary was only worth $9,513. Nevertheless, that $9,513 was more than the $7,858 equivalent starting salary for engineers in 1964. That is, even with a very high rate of inflation during the 15 years from 1965 to 1980, engineering salaries remained constant in real buying power. With decreases in inflation in the mid-1980s, engineers' buying power is increasing because engineering salaries have continued to increase. For example, John Shingleton, Placement Director at Michigan State University, has conducted a Recruiting Trends Survey in which he has investigated the hiring practices of more than 710 employers in business, industry, governmental agencies, and educational institutions throughout the United States. For 1986, the estimated starting salary for bachelor's degree recipients for all fields was expected to be $21,601. This was a gain of approximately 1.8 percent over the 1985 figure. Shingleton reported that demand remained high for minorities and women. Of particular interest is the comparison between engineering salaries and those for other fields given by Shingleton in his survey:

1986 EXPECTED STARTING SALARIES

Electrical Engineering	$29,187.00
Mechanical Engineering	28,971.00
Chemical Engineering	28,739.00
Computer Science	27,775.00
Industrial Engineering	26,817.00
Civil Engineering	24,761.00
Physics	24,370.00
Financial Administration	20,803.00
Accounting	20,338.00
Chemistry	19,679.00
General Business Administration	19,589.00
Marketing and Sales	19,284.00

Mathematics	19,014.00
Social Science	18,324.00
Communications	17,923.00
Agriculture	17,841.00
Advertising	17,832.00
Personnel Administration	17,727.00
Telecommunications	17,473.00
Hotel and Associated Workers	17,375.00
Liberal Arts	17,358.00
Geology	17,185.00
Education	16,903.00
Journalism	16,207.00
Retailing	15,898.00
Natural Resources	15,709.00
Human Ecology and Home Economics	15,635.00

Source: Associated Press, November 26, 1985.

This is the picture for starting salaries. What about long-term salaries? The accumulation of experience in a particular engineering field leads to much higher salaries. Several recent studies of engineering salaries indicated that engineers with 20 years' experience earn approximately 70 percent more than recent graduates, if they are employed in non-supervisory capacities. For engineers who are employed as supervisors, the accumulation of 20 years' experience, on the average, leads to salaries more than double those of recent engineering graduates. A number of studies have been conducted in the last 15 years to investigate the relative positions of engineers' salaries with respect to those of other professional and non-professional people. These studies have shown that engineers and other technical professionals have moved up in income level when compared to other professionals, farmers, owners of small businesses, and blue collar workers. This is particularly true for engineers with significant experience. Engineers now constitute approximately 20 percent of the top income level group in the United States. Civil engineering salaries will be described in more detail in Chapter 4.

MINORITIES AND WOMEN

The role of women and minorities in engineering in the future is likely to expand greatly for two reasons. First, at the present time the enrollment of women and minorities in engineering is far below what it could be and what the demand for women and minorities in engineering jobs would justify. Secondly, the demand for women and minorities in engineering jobs is likely to increase at a higher percentage rate than the overall increase in the number of engineering jobs for all groups. Women, blacks, Hispanics, and American Indians are all underrepresented in engineering. Women accounted for only about 16 percent of the total undergraduate engineering enrollment in the United States in 1984, while representing more than half of the total population. Black undergraduate enrollment also was grossly out of proportion to the percentage of black citizens as a fraction of the total population of the United States. This underutilization of women and minorities in engineering is doubly ironic since a significant amount of the total number of graduate degrees at the master's and doctoral level given in the United States each year are given to foreign nationals. Citizens of countries all over the world are coming to the United States to receive graduate engineering degrees and going forth to practice engineering, competing for positions that could be sought by women and minorities as well as white male Americans. This situation has been very slow to change in the United States, despite vigorous recruiting efforts by engineering societies, colleges, and universities. At present, starting salaries in many fields are higher for women graduates and minority group members than for white male Americans. Despite this obvious advantage, enrollment by women and minorities in engineering programs has not increased significantly in recent years. The conclusion that can be drawn from this situation is that for those women and minority group members who *do* choose to go into engineering, the job market will be extremely attractive in the 1980s and 1990s.

CONTRIBUTIONS—NOW AND IN THE FUTURE

The descriptions of the activities of engineers, and particularly of civil engineers, given in the preceding pages indicated that civil engineers can contribute very importantly to the improvement and maintenance of the quality of life in today's complex technological society. Because of the great expansion of scientific knowledge in recent years, the basis for applying science has been greatly expanded. Also, much non-scientific knowledge has been gained by experience in many different applications of technology. Civil engineers use scientific principles and their own accumulated experience in conceiving plans and designs for physical systems and facilities in a wide variety of ways. Consider, for example, these ads for engineering positions which have appeared in issues of *Civil Engineering* magazine, published by the American Society of Civil Engineers:

Dames & Moore, an international consulting, engineering, and geoscience firm, has immediate openings in its California offices for entry-level and experienced professionals in the following disciplines: Civil Engineers, Geologists . . .

Sr. Civil Engineer—BRPH, one of Florida's fastest growing A/E firms has an outstanding opportunity for a senior design engineer. We seek a degreed and registered professional engineer with 5-7 years A/E experience. Florida registration preferred.

PROGRAM MANAGER—BASIC RESEARCH IN CIVIL ENGINEERING (GEOTECHNICAL). The Air Force Office of Scientific Research, Air Force Systems Command . . . invites applications from qualified U.S. citizens with a background in university or industrial civil engineering fundamental research in geotechnical engineering (soils, structural materials, structures) . . . The position is career Civil Services GM-13 ($37,599-$48,376 per year) or GM-14 ($44,430-$57,759 per year) depending upon qualifications.

[February 1985 issue]

Material and Geotechnical engineers. Pittsburgh Testing Laboratory is a national multi-discipline engineering testing firm which is seeking growth-oriented professionals in geotechnical and materials engineering. Positions exist throughout the United States for professionals with both engineering and managerial backgrounds.

Construction/engineers. Tax free income. Africa—Orient—Egypt—Mid-East—Far East—Turkey. Over 11,500+ openings. Free travel, free housing, free food and free medical. Overseas Unlimited Agency.

[June 1985 issue]

Technical Director ($68,700 per annum). The U.S. Army Cold Regions Research and Engineering Laboratory (USACRREL), Hanover, NH, is recruiting to fill the senior executive service position of technical director...

Overseas employment. Engineers and technicians are needed as instructors for U.S. AID-FUNDED PROJECT IN ZIMBABWE. Recent applied field experience essential. 2 years + contract.

Civil engineers: G & O offers continuous careers in transportation engineering. Here is your opportunity to join Greenhorne and O'Mara Inc., a 35-year-old engineering and design leader...

Combine business with pleasure. Live and work in Maine. Senior structural engineer. Join a growing civil/environmental consulting firm located on the coast of Maine less than one hour from lakes and ski areas...

Civil engineers needed. Arizona Department of Transportation. A landmark highway construction funding bill was passed by this year's legislature setting the stage over the next 20 years for a 6 billion dollar expansion of the state highway program...

[November 1985 issue]

Wastewater water treatment plant O & M specialist. Prominent consulting environmental engineering firm seeks wastewater and/or water treatment plant O & M specialist...

Civil, structural and traffic engineers. Civil and Mechanical designers and drafters. ENR top 500 design firm in sunny Las Vegas,

Nevada can offer you top pay, benefits and advancement
opportunities...

[December 1985 issue]

These advertisements for engineering positions indicate the wide
range in activities open to civil engineers today. In formulating designs and in solving practical problems, these engineers will be
seeking to use available resources in the most efficient and effective
way. The natural resources which are available to engineers include
energy, materials, and human resources. Energy is obtained directly
from the sun, from fossil fuels, from nuclear reactions, and from
other sources such as hydroelectric power and wind power. Some of
these forms of energy are renewable; others are not. Civil engineers
are vitally concerned in energy generation and transmission, working with electrical engineers to provide convenient available power.
Civil engineers also are prime designers in hydroelectric power in-
stallations such as dams. In all of these efforts, civil engineers will
attempt to use nonrenewable resources in the most efficient way possible and to use, wherever possible, renewable fuels rather than those
which are limited and cannot be replaced

These comments reflect the fact that engineers must be aware of
the social and political conditions which have a bearing on their
work. In these examples, it can be seen that the scarcity of energy resources is a prime issue. Certain fuels and raw materials are found
only in limited areas on the earth. The people who live in these areas
may not be willing to share their resources with others. In this most
fundamental way, a social and political restraint is imposed on the
work of engineers. Other social and political factors also enter into
the solution of other technical problems. If consideration is not given
to social and political factors, the solution that the engineer develops
is unlikely to be workable in the real world. This consideration is particularly important to civil engineers because they often deal with
public works projects and come into direct contact with the people
they serve. Obviously, these engineers must not be ignorant of the social and political implications of their activities.

Finally, in recent years, an awareness has developed of the fragile
relationship between human beings and the environment. There is a

growing realization of the limits of the earth with respect to natural resources and the purity of the atmosphere, the waters of the earth, and the land itself. An engineer can no longer simply identify a demand for a particular product or service by society and then attempt to satisfy that demand without considering the consequences of the action. The engineer must decide if satisfying that particular need will lead to a general benefit or will lead to an overall decrease in the quality of life for everyone who could be affected. The engineer tries to consider alternative ways to accomplish the same purpose. In considering these alternatives, social and aesthetic effects also are evaluated. For example, building a dam to create a lake along the valley of a mountain stream may produce a definite benefit in the form of the power supplied through turbines installed in the dam. Boaters and others who want to use the lake as a recreational resource also would be considerably benefited. However, construction of the dam could also rob those same people of an important recreational area in terms of the unspoiled valley of the river. Construction of the dam could eliminate a wilderness habitat important to many other forms of life. An engineer planning and designing such a facility obviously would have to consider the total effects and impacts of any proposed construction at the site. In a nutshell, civil engineers must be aware of the consequences of their actions—they have obligations.

OBLIGATIONS

In evaluating the potential effects of a course of action that is being considered, the engineer obviously must use her or his own personal judgment. In many instances, the choice that the engineer makes will be very simple and will essentially be a matter of choosing a product or a process with the lowest cost which will accomplish a desired objective. In other situations, the engineer's choice will be much more difficult. In some cases, the engineer will be faced with a dilemma in trying to determine the value of a particular asset. Sometimes the asset is intangible, such as a scenic vista down a valley to the sea. It will be difficult for the engineer to try and fix a quantitative value on the cost of preserving that scene and not constructing a highway through the middle of the valley, or constructing the highway in order to gain the benefits of good transportation, but at the cost of sacrificing a beautiful landscape. In making these choices, the engineer is not making a moral decision or deciding a question of ethics. He or she is not trying to decide whether or not something is right or wrong, but rather to develop some measure of the worth of unlike things in order to choose between two or more alternative courses of action which will affect the conditions or things to be evaluated.

In other cases, the engineer is faced with problems which do not necessarily call for an exercise of professional judgment, but demand a moral judgment. These questions of ethics obviously are decided on a personal basis, on the individual's standards of right and wrong. This type of decision is very difficult in many instances, particularly for young, inexperienced engineers. To serve as a general guideline for decisions on ethics and to assist engineers in judging the effects of their activities, the Engineers' Council for Professional Development has developed a Code of Ethics.

Code of Ethics of Engineers

FUNDAMENTAL PRINCIPLES

Engineers uphold and advance the integrity, honor, and dignity of the engineering profession by:

I. using their knowledge and skill for the enhancement of human welfare;

II. being honest and impartial, and serving with fidelity the public, their employers, and clients;

III. striving to increase the competence and prestige of the engineering profession; and

IV. supporting the professional and technical societies of their disciplines.

FUNDAMENTAL CANONS

1. Engineers shall hold paramount the safety, health, and welfare of the public in the performance of their professional duties.

2. Engineers shall perform services only in the areas of their competence.

3. Engineers shall issue public statements only in an objective and truthful manner.
4. Engineers shall act in professional matters for each employer or client as faithful agents or trustees, and shall avoid conflicts of interest.
5. Engineers shall build their professional reputation on the merit of their services and shall not compete unfairly with others.
6. Engineers shall act in such a manner as to uphold and enhance the honor, integrity, and dignity of the profession.
7. Engineers shall continue their professional development throughout their careers and shall provide opportunities for the professional development of those engineers under their supervision.

(Approved by the Board of Directors of the Engineers' Council for Professional Development, October 5, 1977.)

Members of the engineering profession obviously will be called upon to make difficult decisions on the basis of an ethical framework which is a very individual matter. Many other decisions that the engineer makes in the course of routine, day-to-day practice will not require moral judgments but will simply be technical evaluations of factors and situations which can be expressed easily in numerical terms. A truly successful civil engineering practitioner should be an individual of excellent character, and obviously must be technically competent. However, in order to be really successful an individual must have other qualities and characteristics besides integrity and technical ability.

THE IMPACT OF COMPUTERS

The development and widespread application of digital computers in the last twenty years has revolutionized civil engineering, as well as all other branches of engineering. The use of solid-state devices has permitted the invention of miniature equipment so that computers now are small and portable. To equal the computing power of a portable microcomputer that you can hold on your lap today would have required more equipment in 1965 than you could put into the room in which you are sitting. Computers store information as magnetic impulses on a conductive material base. Information can be transferred and processed in a computer at the speed of electricity. Thus, computers allow people to store tremendous amounts of information in very compact form, and to do operations on that information extremely rapidly. Almost all the uses of computers in engineering are based upon the computers' capacity for information storage and retrieval, and the speed of information processing possible with a computer. Today's civil engineer spends a considerable amount of time developing computer programs. However, after these programs, or sets of instructions, are developed, the engineers then are capable of designing and analyzing even the largest structures in a matter of only minutes. Since the calculations performed by use of the computer are carried out electronically, millions of calculations are possible each minute. This is the type of problem to which engineers first applied computers almost exclusively.

With today's computers, engineers are able to take a relationship which describes how one factor depends upon another factor, and approximate even a widely varying relationship as a number (a very great number) of very short line segments. The relationship between one factor and another thus is called linear in the short interval between the subdivisions, and the calculations describing the relationship of one factor to another are consequently very simple. This ability to develop accurate approximate solutions is one of the most powerful tools developed from the use of computers, which have changed the way engineers analyze real systems today.

An entire book could be written describing all of the uses to which computers are placed today by civil engineers. By the time the book was written a multitude of new uses would have been developed. Our purpose here is not to describe all of the ways in which engineers use computers but simply to show you a number of the more important ways in which computers have assisted civil engineers in the type of work that they have been carrying out for many years. The computers do not take the place of judgment and insight developed by the engineer. However, the engineer now is able to carry out many more calculations than were ever possible in the past, and can perform analyses by methods which were never sufficiently accurate in the past, but which now, with the use of computers, are extremely accurate. In the future, many more applications of computers in civil engineering will be made. Perhaps you will develop one of these new techniques. Maybe you will even work in the rapidly expanding field of artificial intelligence/expert systems.

ARTIFICIAL INTELLIGENCE AND EXPERT SYSTEMS

Artificial intelligence, or AI as it is known, is an area that offers an unlimited potential for use by civil engineers. What is AI? It is development and application of computer hardware and software that simulates human thought and action. Specific topics within AI that are of interest to civil engineers include robotics, image processing, and, most important, expert systems.

Expert systems are rapidly becoming an integral part of advanced civil engineering activity. Basically, an expert system is an "intelligent" computer program that uses knowledge and inference techniques to solve problems difficult enough to require significant expertise. An expert system is different from a conventional computer program in that it contains several separate components modeled after the way human experts think and make decisions. These components include a "knowledge base," which is a collection of known facts and less well-defined rules of thumb; an "inference engine," to process the facts and rules of thumb; and a "work space" that maintains the status and data for the problem being solved.

This definition of an expert system could be confusing, and its applicability to civil engineering may be unclear. Therefore, let us consider a sample problem with a civil engineering slant. Actually, we will consider only a small, extremely simplistic piece of a typical civil engineering situation.

Before we look at this problem, however, we first need to discuss the computer languages commonly used in expert systems. Unlike conventional algorithmic programs which are usually written in FORTRAN, PASCAL, or BASIC, most expert systems are written in

LISP or PROLOG. These two languages allow natural English language interaction with the computer, and are designed to process lists of data, rather than mathematical relationships.

Our problem is to determine the number of lanes a highway entrance into a shopping center needs to have in order to meet traffic capacity requirements, but not be overdesigned and needlessly expensive. As a highway designer, you have access to the following facts:

1) one lane of highway can accommodate 1,500 vehicles per hour;
2) for every two percent of the traffic that is trucks, total capacity is reduced three percent; and
3) the maximum speed allowed in the area of the shopping center is 20 miles per hour.

As an expert highway designer with years of experience, you also have developed the following rules of thumb to aid you in design:

a) shopping center traffic usually travels significantly more slowly than posted speed limits allow; and
b) unless the entrance road is at least 1,000 feet long, its actual capacity will only be about 80 percent of the ideal.

To construct an expert system to solve this problem, these facts and rules of thumb are programmed as "rules" into the knowledge base in a list which is examined in a sequential manner by the inference engine. A typical rule would actually look like this in computer code:

```
IF max_vehicles_per_hour GT 1500
AND max_vehicles_per_hour LT 3000
THEN add_one_lane
```

where "GT" means "greater than" and "LT" means "less than."

The inference engine portion of the expert system is computer code which is written to execute the IF-THEN rules in the same way as the human mind does. For example, the human expert would mentally deal with the following information in more or less this order:

```
IF the highway will carry 1,900 vehicles per hour, and
IF it must accommodate 20 percent trucks, and
IF the speed limit is 20 miles per hour,
```

THEN the highway must have three lanes.

Just like the human mind, the inference engine would obtain the solution by comparing the given data to each of the facts and rules of thumb in the knowledge base, making the necessary calculations, and noting any violation of any stated parameter. The outcome, three lanes, would then be evaluated against each rule one more time to make certain that none is "fired," or invoked. When no more rules are fired, the expert system informs the user that the correct number of lanes is (for this case) three.

This expert system example may be thought of as simplistic and perhaps even crude, but it does demonstrate a different way of approaching problem solving, as compared to traditional mathematical methods. The power of the expert system lies in its extraordinary interactive capabilities, in that the user can change the given input data in almost a conversational interchange with the program (i.e., it is quite user-friendly). Also, the expert system can be easily improved by simply adding new rules to the list in the knowledge base.

Civil engineers are among the first to embrace expert systems as a tool useful in practice. As a result, such programs are now available to assist in building design, geotechnical and subsurface investigations, earthquake analysis, traffic signal setting, pavement management, and so on. Should you choose civil engineering as your profession, you will likely have the opportunity to use and perhaps even develop an expert system as part of your duties.

PERSONAL ATTRIBUTES

If you are trying to decide whether or not you are suited to a career as a civil engineer, you should examine your character and your interests, especially with regard to three areas: involvement, ability, and ambition. In other words, if you hope to be a successful civil engineer, you must be sufficiently interested in the field of civil engineering, you must possess the required capabilities, and you must have sufficient drive and perseverance to succeed. Of these three factors, perhaps the most important is the amount of interest you have in this particular career. You must be somewhat interested in this career or you would not be reading this book at this time. The question you must answer for yourself is how seriously you are interested in civil engineering. Any type of work which greatly interests you will probably be the type of work at which you will succeed. On the other hand, work which you consider boring, uninteresting, or unimportant will be difficult for you to perform. If you are trying to make a mature evaluation of your own interests, you should review your life to see what ideas or activities have aroused your curiosity and attracted your attention. With regard to engineering, you should consider any interest you may have felt in the past in the operation of electrical or mechanical devices, or in technical and scientific developments. Are your hobbies related to scientific and technical activities? Do you like mathematics and mathematical puzzles? When you have read other books or talked to individuals about other career fields, if they talked about engineering or if you read about engineering, did you find the discussions interesting or boring? If you have ever visited a construction project, a manufacturing plant, or an engineering design office, were you interested in the activities of the technical personnel employed in those places? Have you ever asked any engineers, particularly civil engineers, about their work? Have you made an honest effort to learn as much as possible about the field of civil engineering? In other words, are you really serious about this subject? The answers to these questions can go a long way towards indicating to you whether or not you truly have a significant interest in engineering, and in particular, in civil engineering.

In addition to being intrigued and interested in the field of civil engineering, an individual must have certain capabilities in order to succeed. Among these capabilities is an easy mastery of mathematics and of the application of mathematical principles to the solution of practical problems. You should ask yourself: "How good am I at stating problems or describing situations in quantitative terms (using numbers)? Have I had difficulty in obtaining numerical solutions to word problems in my high school classes? Can I look at a real situation and think of it in terms of abstract ideas without becoming confused? Can I discover the underlying principle in the operation of a machine or a device? Do I understand the various number systems that are used in mathematics?" In addition to these questions about mathematics and the application of mathematics, you should think about your interest in related fields such as physics and chemistry. Have you ever tried to apply the principles from physics to analyze a problem in the real world? Can you understand concepts such as force, mass, acceleration, leverage, gravitation, and other abstract physical principles? Are you interested in the way in which various elements combine in nature? Can you see applications of chemical theory in daily life? One of the easy ways to answer these questions, at least partially, is to look back through your high school academic record and review the grades you have obtained in science and mathematics courses. High scores and excellent grades in mathematics, physics, and chemistry usually indicate that a person has the capability and interest to undertake engineering studies with a significant hope for success.

In addition to these academic capabilities, some other characteristics are also necessary for a successful career in engineering. One of these abilities is the power to visualize in concrete terms what is described to you only in words. If you are interested in a career as an engineer, you should consider whether or not you are able to describe in words, and then to illustrate in sketches, complex actions or processes. For example, would you be able to describe in words and illustrate in a few simple sketches, the mixing of fuel and air in the carburetor of an internal combustion engine? Could you make a simple sketch to show the way in which an airplane wing works to lift the plane off the ground? The ability to communicate, on paper and orally, is very important to your success as a civil engineer. You should also consider whether or not you get along well with other people. No engineer works alone or in a vacuum. An engineer working as a member of a large design team or a large construction force must be able to get along well with other people and communicate effectively with them. You should examine your own past performances in working together in groups with other persons to accomplish specific goals and objectives.

Remember that none of the qualifications described above are at all related to sex or racial background. Any person capable of developing abstract concepts, of dealing effectively in quantitative terms with physical processes, and of communicating effectively with others can succeed in a career as an engineer. There is no reason to expect that members of minority groups would have any trouble in obtaining interesting employment at very attractive salaries in the field of civil engineering in the immediate future and for a long time to come. Women and minorities will find ample opportunity for employment and advancement. This advancement will occur, if the individual performs as expected, whether or not that person is employed by a federal or state government or by private industry. A person's capability to produce practical solutions to real problems is much more important than race, sex, or ethnic background.

In making the self-evaluation called for in the preceding paragraphs, if you have been honest and have judged correctly your interests, abilities, and ambitions, you can say at this point whether or not you are seriously interested in civil engineering as a career. You must remember, however, that you will be required to work hard and persevere, at times under very unfavorable conditions. Working in technical fields can be difficult and very exacting. As a civil engineer, you will be called upon to master a great deal of technical information and to understand many technical concepts. You must be capable of applying yourself in an efficient way. You should reexamine yourself constantly to improve your efforts. If you can work in this way, and if you are sufficiently ambitious, you will succeed in a civil engineering career just as you would in any other career field. On the other hand, a careless or lazy individual can no more succeed as a civil engineer than in any other profession. In order to get an estimate of your own capabilities in this regard, ask yourself the following questions: "Do I complete my study assignments on time? Do I need urging by others to complete an assigned task? When I have an assignment, do I avoid the technical or numerical portions of a problem? Am I more strongly drawn toward non-technical questions? In general, do I get the job done or am I usually late in completing assignments?" The answers to these questions will help you to judge your potential for success in a career as a civil engineer.

A group of foreign engineers tours the powerhouse at McNary dam on the Columbia River. (Army Corps of Engineers photo)

CHAPTER 2

EDUCATION FOR CIVIL ENGINEERS

If the description of civil engineering which was given in Chapter 1 is interesting to you, you should begin to consider a number of questions. What should you study in high school if you want to attend engineering school? How should you go about deciding which college or university is best for you? What will you study if you do enter an engineering school? If you choose a career in civil engineering, what types of programs are available? In this chapter, we will attempt to answer these questions and others to help you learn as much as possible about the field of civil engineering and the education which is necessary in order to practice as a civil engineer.

Obviously, the education of a civil engineer begins in grade school with the study of such elementary subjects as English and arithmetic. In reality, a civil engineer's education ends only after retirement or after leaving practice as an engineer. Between these two end points, engineering education moves from a study program with emphasis on the details of technology and the sciences through the gaining of years of experience, with emphasis on judgment and management rather than scientific fundamentals. Compared to the entire period in which a professional engineer acquires experience and applies knowledge in practice, the four to seven years of formal education in a college or university are only a brief interlude in her or his education.

Formal study in civil engineering can be considered to have several goals. A good curriculum should be well rounded; that is, it should prepare the student for productive practice in all aspects of the civil engineering profession. Obviously, formal education should provide the student with sufficient mathematical, scientific, and non-technical background to accomplish fundamental technical tasks. Also the civil engineering student should receive sufficient formal education to permit her or him to continue self-education throughout an entire career. A formal course of study should introduce engineering students to the social, political, and economic sciences so that in the future, as civil engineers, they will be capable of working with specialists in those areas, and of developing engineering designs which will take into account multiple aspects of problems. In other words, the engineer must be able to consider social, political, and economic consequences of designs and plans as well as the intended technical results. Finally, an engineering curriculum must provide adequate preparation for those exceptionally capable students who wish to continue their education beyond the first professional engineering degree. Ultimately, this formal education in college is based upon what the student has learned in high school.

HIGH SCHOOL

The rapid pace of change in today's technological society has already been described. New knowledge in all fields of engineering, as well as in the particular field of civil engineering, is becoming available at an astonishing rate. Engineering education, in order to cope with these rapid changes, has become more complex and more difficult. New engineering graduates must possess a very sound background in the physical sciences, in mathematics, and in communication skills. This background in turn must be based upon high school preparation in these areas. A thorough grounding in English communication, mathematics, and physical sciences will be required for entrance into almost every engineering program in the United States. Engineers are required to communicate with other persons, including technologists and non-engineers as well as members of the general public. It is imperative that engineers have a good command of written and spoken language. If an engineer cannot describe designs and plans, if he or she cannot put the solution to a problem into words, that solution will not be implemented. In other words, it will be no solution. Most students would recognize that it is necessary to have a thorough grounding in mathematics and in physical sciences, but it is also important to have a solid background in language skills.

Since it is obvious that you should have a thorough grounding in these curriculum areas in high school, it is equally obvious that you should begin as early as possible to develop your educational plans to satisfy the requirements for engineering school and for an engineering career. Unfortunately, many students do not decide upon a career until they are well along in their high school education. Some have not decided upon a career at the time of their graduation. As a result, many students attempting to enter college or university engineering schools find themselves lacking the necessary requirements for admission. A number of these students are required to spend additional time taking makeup courses before they are formally admitted to engineering school. If you want to avoid this, you must give serious thought to your career plans as early as possible and begin taking those courses which will be required for admission into the professional program you select. Fortunately, the courses which are required for entrance into engineering school are also very important courses as background to many other careers. In other words, if you have not completely decided to pursue a career in engineering, but you take those courses necessary for entrance into engineering school, those same courses will be valuable background if you choose a career in any of the sciences, and they will not represent lost time for a wide variety of career choices.

While the exact requirements for admission to engineering schools in various universities differ from school to school, these differences usually will not be significant. In general, the requirements for entry into engineering school will be as follows: a total of 15 or more standard high school units of credit, of which 12 must be academic, not vocational in character. The distribution of these 12 units should be approximately as follows: English, 4; algebra, 2; plane geometry, 1;

advanced mathematics (trigonometry, introduction to analytic geometry, calculus or equivalent), 1; chemistry, 1; physics, 1; the remainder may be history, social studies or foreign languages. Some basic knowledge of mechanical drawing and drafting also would be helpful in several branches of engineering. If possible, other physical science courses should be included in the high school program.

In addition to these courses in English, mathematics, and the sciences, it would be advantageous for you to take a course in computer programming if one is available in your high school. Having a course in typing would be similarly advantageous. Both of these types of courses would be extremely useful from a practical viewpoint in engineering school, and in most other curricula. In some cases, typing courses have been converted to use computer-based word processors so that it is possible to get some experience with the computer hardware at the same time you are learning typing skills. If you do not have a chance to take these courses, you will not be seriously at a disadvantage in college. If you can take them, however, without interfering with the other more important courses such as English, math, and science, they can be a bonus for you in engineering school.

In addition to planning your high school program to include those courses necessary for entry into engineering school, you can also prepare for a career in engineering in other ways. For example, you can investigate hobbies and crafts that are associated with technical activities. Do not misunderstand; it is not necessary to be highly skilled in a wide variety of crafts or to have a high degree of manual dexterity. You should simply seek to develop some familiarity with tools and instruments, and try to develop a certain degree of mechanical skill. Joining a student association or club devoted to a technical hobby also can be extremely useful in indicating to you what abilities you would have in these technical areas. This kind of activity has an added advantage in that it will indicate to you how deep is your interest in these technical areas. Participation in various hobbies which are related to science and engineering will not only offer an informal education in these areas which can be valuable in later years, but it can also be very useful in telling you whether or not you are really interested in engineering and science as a career.

Finally, during high school it is important to acquire good study habits. Engineering students find themselves pressed for time to complete all of their assignments and keep current with all of their courses. This is true for students who had little difficulty in high school, as well as for those who had to work hard. There is a great deal of technical material which must be mastered in an engineering curriculum. Many engineering courses require two to three hours of study outside the classroom for each hour the student is in the classroom. Also, students who were at the top of their graduating class in high school may suddenly find in engineering school that they have a number of classmates who are their intellectual equals, if not superiors. It will not be possible to "coast" through engineering school on the basis of inborn ability and talent. Long hours of study will be required from *everyone.* Budgeting of study and leisure time is an absolute necessity. If the student can acquire these good study habits during high school, this investment will pay handsome dividends many times over in college and later in professional life.

CHOOSING A COLLEGE OR UNIVERSITY

If you think that a career in engineering is really for you, the next important decision you face is a choice of college or university to attend. In this chapter, we will explain in detail the various types of engineering education programs that are available today in the United States. Before you look at those programs, we would like to give you some tips on how to evaluate universities and their engineering schools.

Basically there are four major points to consider in choosing a college or university. These points include the university faculty, the university physical plant and facilities, the students, and the curriculum. All of these factors are important in creating the kind of environment in which students may be adequately prepared for future practice in the profession. The faculty should be well-rounded and diversified in their backgrounds and educational experiences. This diversity is important if the faculty are going to give the undergraduate students a comprehensive preparation for the engineering profession. When you are examining a brochure or catalog from a university, look at the number and type of advanced degrees each of the faculty members holds. Look at the schools from which these degrees were obtained. In general, a faculty holding advanced degrees from a large number of outstanding schools will provide an excellent instructional environment.

Obviously, an adequate physical plant (the classrooms, libraries, dormitories, and other physical facilities of a university) is necessary, particularly today when engineering laboratories must be equipped with the most sophisticated devices and apparatus. University computer resources, particularly resources available to the undergraduate student, also must be considered. In addition, other facilities also are necessary if you are to obtain a well-rounded education. Consider the facilities for extra-curricular activities such as intramural sports, student publications, and other activities. Look at the dining facilities and the recreational facilities on campus available to all students. Consider the entire physical setup in evaluating the facilities.

Another important feature of the educational environment in a college is often forgotten by the public. This important factor is the student body itself. Generally, a group of students with diverse backgrounds, both geographic and economic, and with a reasonable number of international students, will provide a broader educational experience for the student than will a student body whose members all come from the same geographic area or from similar economic backgrounds. A college education should be a total learning experience. The time spent in a university setting often is a crucial time in a person's entire life. Much of the learning which takes place on a university campus is learning from peers. When you evaluate a school, try to find out something about the makeup of the student body, and if possible, talk to some of those students yourself.

Of course, one of the most important factors to consider in making your choice is the engineering curriculum available at the particular school you are examining. With the explosion of engineering knowledge and the rapid pace of technological change in civil engineering today, it is no longer possible for any one curriculum to offer all things to all people who wish to enter civil engineering. In almost every curriculum, it is necessary to specialize to some extent in one or several of the areas of practice in the field of civil engineering. These areas of specialization will be explained in detail in Chapter 3. Every school will require its students to complete a minimum number of courses which form the basis for the civil engineering curriculum. Generally, this core consists of about 30 percent of the entire study program. The remainder of the curriculum usually is divided among the humanities, engineering courses, and science courses relevant to the student's area of specialization. Look carefully at the curriculum of the school you are considering to determine if the areas of civil engineering specialization offered at that school include the specialization that interests you as a future career.

After you have selected one or more universities for further investigation, your next step will be to obtain information regarding the specific details of school programs and the requirements for admission. Almost all accredited colleges will require a transcript of high school courses you have taken and grades you have received. These schools will also require that you submit scores from college entrance examinations. Most schools will accept either the American College Test (ACT) or the Scholastic Aptitude Test (SAT) as an indicator of your academic capabilities. In addition, some colleges require letters of recommendation, their own particular entrance examinations, medical examinations, and perhaps an interview with an admissions officer before you are considered for admission. You should get the specific entrance requirements directly from the college or university in which you are interested. This information can easily be obtained by writing to the admissions office of the school you are considering and requesting a catalog, an admissions application, and copies of any brochures that the school may have which describe its civil engineering program. Do not put off obtaining this information until your senior year in high school. As we have already pointed out, it may be necessary to assemble a number of different materials and take specific examinations given in various localities only at certain times during the year. In addition, there may be a time delay in obtaining letters of recommendation necessary to support your entrance application. For these reasons, it is a good idea to begin as early as possible to investigate the admission requirements of the various colleges you are considering.

For those who are already enrolled in a college program and who are considering transferring into a program in civil engineering, it is usually possible to make such a transfer. In many cases, a number of credits earned during previous college work can be transferred to count against the requirements for an engineering degree. However, each school has its own particular policy on the transfer of credits from one institution to another, or from one program to another within a given university. You should contact the admissions officer or registrar of the school you wish to attend to determine specific procedures for transfer to that school.

Finally, it is safe to say that you can never have too much information. Try to discuss with alumni from a given school all of the aspects of going to that particular institution. Many times, graduates will be able to offer you valuable insights that you would not find included in any of the publications from a university. If you can, discuss the college or university with one or more of the students who are currently attending that school. With regard to a civil engineering curriculum, you can also talk about that school and that program with practicing engineers. Their opinions should be considered, along with all the other information you have gathered, before you make your final decision.

CURRICULA

Many variations in civil engineering curricula have come into being to meet the needs of a wide variety of students. Some students want to go into civil engineering practice at the end of a bachelor's degree program, while others want to continue working towards an advanced degree. Additionally, some students want to emphasize one particular facet of civil engineering more than any other. Finally, some students want to obtain some practical experience while they are in school. There are programs available today in the United States to satisfy all of these different needs. Generally, a student will take a core of science and basic engineering subjects which will be common to all undergraduate engineering programs. Then the student will take a similar core of civil engineering subjects, selected by the faculty members at the particular school as representing those courses necessary for every civil engineering student to take. After the student finishes this core of civil engineering subjects, he or she can choose to concentrate the remainder of the undergraduate courses in a specialty area, or he or she may distribute the courses to cover a number of different aspects of civil engineering. The flexibility and latitude of choice allowed to the individual student varies from one school to another.

There are almost as many varieties of civil engineering programs available for the prospective student to choose from as there are civil engineering schools. Some of the courses of study will allow relatively little choice on the part of the student, while other schools allow the student wide discretion in making course selections. At some schools, all civil engineering students will follow very similar programs, while at other schools there will be one or more options available for specialization by the student. At some schools, the entire college period will be spent in academic work, while at other schools, students will gain professional working experience through programs that involve cooperative internship. All of this variety in programs makes your choice of a college or university more difficult. To illustrate the variety of choices which are available, in the following pages we have attempted to give examples of a few of the types of curricula that are available at present. All of the programs described in the following pages are fully accredited by the Accreditation Board for Engineering and Technology and thus can be considered good civil engineering programs.

BACHELOR'S DEGREE PROGRAMS

The traditional formal education of civil engineering students has been a four-year program, relatively inflexible, leading to a bachelor's degree. By far the vast majority of engineering schools in the United States today still retain this format for the first accredited degree. The basic course requirements of most of the schools are quite similar. The curriculum usually consists of approximately one-fourth mathematics, physics, and chemistry; one-fourth English, humanities, and social sciences; one-fourth engineering theory; and one-fourth engineering analysis and design. The specific courses available at any given school have been selected by the faculty of that school to fulfill the requirements of the Accreditation Board for Engineering and Technology. There will be minor differences from one curriculum to another in the courses that are required, as well as the number of electives that the student is permitted to choose. Also there will be minor differences in the course content in the individual courses. In this type of program, however, the electives will not be numerous and the required courses usually will differ only in the depth of study into one or more of the civil engineering specialized areas. Typical of this type of program are those shown below:

What Do Bachelor of Civil Engineering Majors Study?

The civil engineering curriculum is characterized by a broad range of coursework which prepares the graduate not only to function technically within the civil/aerospace communities, but also to relate to other engineering disciplines and to nontechnical components of society. The freshman and sophomore years are devoted largely to building a sound base of knowledge in mathematics, physics, chemistry, and elementary engineering science and mechanics. The junior and senior years focus on technical subjects related primarily to civil engineering, with a number of engineering electives available in the senior year to permit a student either to specialize in a particular field of interest or to prepare for entrance into a graduate program of study. Interspersed throughout the curriculum is a series of courses of a general nature which provides students of civil engineering with sound philosophical, historical, and social foundations, which prepares students to communicate both orally and in writing, and which instills an appreciation of the arts. Here is a sample course sequence from the University of Dayton:

BACHELOR OF CIVIL ENGINEERING

Freshman Year

Engineering and Scientific
  Programming
General Chemistry I
Statics
Introduction to Engineering
English I
Analytic Geometry and Calculus I, II
Engineering Design Graphics I
General Physics I
History of Western Civilization
Introduction to Philosophy

Sophomore Year

Analytic Geometry and Calculus III
General Physics II, III
Surveying
English II
Strength of Materials
Engineering Graphics II
Applied Differential Equations
Highway Geometrics
Fundamentals of Effective Speaking
Engineering Geology
Dynamics
Seminar
Surveying Field Practice

Junior Year

Hydraulics
Analysis of Determinate
  Structures
Civil Engineering Analysis
General Chemistry II
Philosophy or Religion Elective
Soil Mechanics
Analysis of Indeterminate Structures
Civil Engineering Laboratory
Sanitary Engineering
Social Studies Elective
Arts Elective
Seminar

Senior Year

Design of Steel Structures
Environmental Engineering
Transportation Engineering
Engineering Ethics
Science/Engineering Elective
Design of Concrete Structures
Civil Engineering Elective
Civil Engineering Elective
Philosophy or Religion Elective
History Elective
Seminar

OPTIONS WITHIN BASIC PROGRAMS

There are many separate areas of specialization in civil engineering and many schools offer the opportunity to specialize in one of these areas. In other words, the student can select an optional set of courses concentrating in one specialty area, leading to the bachelor's degree. In contrast to the programs shown in the previous section in which a limited number of electives could be chosen, in this type of curriculum two or more sequences of courses could be selected to give in-depth coverage in a particular area of specialization. Among the areas of specialization commonly found in the engineering schools in the United States are water resources engineering, structural engineering, transportation engineering, public works engineering, construction engineering, and geotechnical engineering. All of these areas of specialization are described in more detail in Chapter 3. The following programs are offered at North Carolina State University. The first is the basic civil engineering curriculum, and the second is the specialized construction engineering curriculum.

CIVIL ENGINEERING CURRICULUM

Freshman Year

Fall Semester

General Chemistry
Introduction to Engineering *or*
  Engineering Graphics
Composition and Rhetoric
Analytic Geometry & Calculus I
Humanities & Social Science
  Elective
Physical Education

Spring Semester

Chemistry—Principles &
  Applications
Engineering Graphics *or* Introduction
  to Engineering
Composition & Reading
Analytic Geometry & Calculus II
General Physics
Physical Education

Sophomore Year

Fall Semester

Intro. to Civil Engineering
Engineering Mechanics—Statics
Geometry & Calculus III
General Physics
Engineering Economic Analysis
Physical Education

Spring Semester

Physical Geology
Mechanical Properties of
  Structural Materials
Applied Differential Equations I
Mechanics of Solids
Engineering Mechanics—Dynamics
Humanities or Social Science
  Elective
Physical Education

Junior Year

Fall Semester

Engineering Surveying
Structural Analysis
Materials of Construction
Hydraulics
Humanities or Social Science
  Elective

Spring Semester

Transportation Engineering I
Structural Engineering I
Soils Engineering I
Water Resources Engineering I

Senior Year

Fall Semester

Two of the following four:
    Transportation Engineering II *or*
    Structural Engineering II *or*
    Soils Engineering II *or*
    Water Resources Engineering II
Engineering Science Elective
Math or Statistics Elective
Humanities or Social Science
  Elective
Free Elective

Spring Semester

Civil Engineering Design
Civil Engineering Elective
Free Electives
Humanities or Social Science
  Elective

CONSTRUCTION OPTION CURRICULUM

Freshman Year

Fall Semester

General Chemistry
Introduction to Engineering *or* Engineering Graphics
Composition and Rhetoric
Analytic Geometry & Calculus I
Humanities & Social Science Elective
Physical Education

Spring Semester

Chemistry—Principles & Applications
Engineering Graphics *or* Introduction to Engineering
Composition & Reading
Analytic Geometry & Calculus II
General Physics
Physical Education

Sophomore Year

Fall Semester

Intro. to Civil Engineering
Engineering Mechanics—Statics
Geometry & Calculus III
General Physics
Engineering Economic Analysis
Physical Education

Spring Semester

Physical Geology
Mechanical Properties of Structural
  Materials
Applied Differential Equations I
Mechanics of Solids
Engineering Mechanics—Dynamics
Humanities or Social Science Elective
Physical Education

Junior Year

Fall Semester

Engineering Surveying
Structural Analysis
Materials of Construction
Hydraulics
Humanities or Social Science Elective

Spring Semester

Transportation Engineering I *or*
    Water Resources Engineering I
  Structural Engineering I
Soils Engineering I
Construction Engineering I

Senior Year

Fall Semester

Cost Analysis & Control
Construction Engineering II
Engineering Science Elective
Free Elective
Humanities or Social Science Elective
Mathematics or Statistics Elective

Spring Semester

Civil Engineering Project
Legal Aspects of Construction
Free Electives
Humanities or Social Science Elective

ADVANCED STUDY IN ENGINEERING

Many engineers presently in practice feel that the bachelor's degree does not fully prepare a college graduate for a lifetime practice in engineering. Some feel that the bachelor's degree graduate should obtain additional experience before being considered a full-fledged practicing civil engineer. Others feel that more formal education is necessary before becoming a practicing engineer. In all widely recognized professions today (law, medicine, engineering, etc.) additional study, both formal and informal, is advocated as desirable, if not necessary, to obtain the required educational background for practice. Continuing informal education is considered necessary by almost all engineers to maintain competence in the light of the continued technological change which is occurring in the engineering profession.

In the field of civil engineering, the need for more advanced study has been met by three types of programs. The least formal program is that of seminars, short courses, and other short-term meetings. This type of effort is generally called "continuing education." These activities will be described in more detail later in this chapter. The other two types of advanced study are the traditional graduate studies toward advanced degrees such as the Master of Science (M.S.) and

Doctor of Philosophy (Ph.D) degrees, and the professional school, with study directed toward obtaining a Master of Engineering (M.Eng.) degree. These two types of degrees reflect a difference in opinion among engineering educators. This disagreement began during the 1960s, when the American Society of Engineering Education commissioned a study by a group of distinguished educators and engineers. The goal of this study was to define " ... the goals of engineering education" in our rapidly changing society. In its final report, issued in January 1968, this committee recommended more official recognition of the role of graduate study, especially the master's degree program, in the profession. Many people disagreed with the findings of this committee. However, the faculties in a number of schools decided that four-year programs were inadequate for professional engineering education. These groups changed their programs to include a Bachelor of Science degree (not an engineering degree) to be awarded after four years of study, with a professional Master of Engineering degree to be awarded after a fifth year of study. The Master of Engineering degree was considered the first professional degree in these programs.

In the more traditional type of advanced study in civil engineering, the course work was directed more toward research rather than practice as in the Master of Engineering degree. A primary focus of the M.S. and Ph.D. degree programs has been the preparation of the students for work in research or in academically related activities such as teaching. However, as civil engineering education has become more complex and sophisticated, there has been an increasing emphasis on the need for advanced degrees for practicing engineers. The great debate among civil engineers is whether or not the traditional research-oriented degrees, similar to the advanced degrees awarded in the scientific professions, are more suitable for engineering students than the practice-oriented degrees such as the Master of Engineering. In contrast to engineering, in the professions of law and medicine, graduate professional studies are directed almost equally toward practice and research. The emphasis on research in traditional graduate studies in engineering originated in the fact that much of engineering education is scientific in character. Also, when graduate study programs were developed in engineering schools, graduate study programs already existed in the sciences and were well established. For this reason, graduate study programs in engineering often were modeled after graduate studies in science. Different schools have approached the problem of how much emphasis to put on theory, as opposed to how much emphasis to put on practice, in different ways. Some schools have maintained the traditional research-oriented degree programs while other schools offer programs more closely related to actual engineering practice, and some stress a professional orientation in their advanced study programs.

In most cases, the course of studies leading to the master's degree generally consists of a broad civil engineering program taken during one additional year of study, or a concentration of courses in one specific specialization within the general field of civil engineering. In other words, some schools have stressed one additional year of general preparation for civil engineering, while others have used an additional year to focus the student's studies into one area of specialization. Many schools require that the student write a thesis to obtain the Master of Science degree. This thesis must be based upon an original investigation of a problem which can be either theoretical or practical in nature. Some universities will permit the students to substitute additional course work in place of the thesis. In general, most graduate programs are quite flexible and allow the student to select a required number of courses from a diversified group of offerings. Each student works with an advisor and a committee to tailor the selection of courses to her or his particular needs and choice of career specialization. In contrast to the master's degree programs, the course of studies followed for the Doctor of Philosophy degree *must* include the preparation of a written thesis based upon an original research project. The course work involved in the Ph.D. program is similar to that taken for the master's degree, except that the level of study is much more detailed and intense. Additionally, the course work for the doctoral degree usually contains a significant emphasis on research. The preparation for the Ph.D. is designed to prepare the student to become an engineering educator as well as an engineering practitioner. It is virtually impossible to obtain a position as a full-time, permanent faculty member in any engineering school in the

United States without a Ph.D. degree or an equivalent degree. That is not to say that all students who obtain Ph.D. degrees go into teaching. A significant proportion of Ph.D. graduates enter practice but concentrate on professional activities requiring a strong background in research.

Usually, in order to qualify for admission into a master's degree program, the undergraduate student should have a standing in the upper one-fourth of her or his class. If the postgraduate university considers the student's undergraduate university to be rather weak, the student may be required to be in the upper ten percent of the graduation class. However, a graduate school with only moderate standards may admit a student from a strong undergraduate school if he or she stood in the upper three-quarters of the class. Obviously, these requirements will vary depending upon the quality of the school where the student did undergraduate work and the quality of the school where the student wants to pursue graduate studies.

It usually will be necessary to spend at least one academic year pursuing the master's degree, although a student may spend a year-and-a-half or even two years in master's studies if any deficiencies in the undergraduate curriculum must be made up.

If you were to look at the records of most Ph.D. candidates in engineering schools in the United States you would see that they usually rank in the upper five or ten percent of their undergraduate classes. Rarely will a person pursue doctoral degree studies successfully if he or she has ranked below the upper one-fourth of the undergraduate class. Usually a doctoral degree will require a minimum of three years of study after the bachelor's degree. In many cases the time necessary to complete the Ph.D. is longer than this and may be as much as four to six years. One of the reasons doctoral candidates take longer to finish their studies is that usually these students work part-time as teaching assistants or research assistants. The usual engineering doctoral degree is the Doctor of Philosophy (Ph.D.), although some schools offer the Doctor of Engineering (D.Engr.), Doctor of Engineering Science (D.Engr.Sc.), or the Doctor of Science degree (Sc.D.). When you consider graduate studies in civil engineering, you should obtain the bulletins for the graduate schools at the universities in which you are interested and obtain specific details on those graduate programs from those sources.

COOPERATIVE WORK PROGRAMS

Over the years, there have been different approaches to the education of an engineer. In some societies, the educational process for an engineer basically was that of an apprentice who worked under a master engineer in the field and gradually became accepted by the members of the profession after a sufficiently long period of time. In other countries, formal education in engineering schools was required and no emphasis was placed on practical experience. Today in the United States, at a number of universities, programs of study have been established which represent a compromise between these two philosophies of engineering education. At these institutions, students are able to obtain practical engineering experience at the same time they are completing their academic requirements. These programs are known as Cooperative Work Programs. The administrations of these schools work closely with industries and governmental agencies to establish jobs in which the engineering student can gain experience in a meaningful way. Cooperative Work Programs offer the student several advantages, the most important of which is the opportunity to take a realistic look at the world of engineering practice. Such experience will allow the student to evaluate the long term opportunities for work in the engineering profession and to compare the type of work available to the type of work in which he or she is most interested. The student acts as an intern and works with practicing engineers on real problems. This experience also can show the student the variety of specializations available within the broad spectrum of civil engineering. This experience in actual engineering work also is valuable in indicating to the student her or his own capabilities and limitations. The experience of working with practicing engineers on real problems also makes the student's continuing course work more meaningful. When the student graduates, the cooperative work experience makes her or him a more valuable

employee, and allows her or him to command a higher starting salary. Finally, the most obvious advantage of the cooperative work experience is the salary which the student earns on the job; for many students, this cooperative work salary is the primary means by which they are able to pay for the cost of their education.

There are essentially two choices among academic programs which include a cooperative work experience: in some programs, the cooperative internship is compulsory and is a requirement for obtaining the engineering degree for all students; in the other programs, "co-op" is optional and is sometimes available only to the better student. In all cooperative work programs, the overall course of studies necessarily is made longer and, usually, at least five years are required to obtain a bachelor's degree in a program which includes a cooperative work internship.

TRANSFERRING TO OTHER PROFESSIONAL SCHOOLS

Several times we have emphasized that engineering education must include considerable attention to economics and the social sciences, as well as the humanities, since the engineer's work involves close relationships with people in their everyday living. Engineers are educated to investigate problems, discover all of the available information, assemble the complete data including their own experiences, and then to develop conclusions or solutions on the basis of rational processes. All of this is done in a logical and systematic way, which has been found to be useful in other professions as well as engineering. Many people have found that an undergraduate degree in engineering is a valuable background for further study in law, medicine, business, etc. A significant number of students take their undergraduate training in civil engineering and then later transfer to another professional school or to a specialized course of study in another discipline.

Some engineering schools, recognizing this possibility, have developed special options within their curricula to accommodate this type of student. The most important requirement for the student in these types of programs is to recognize sufficiently early in the studies that he or she really wants to go on to a different program at the graduate level or that he or she wants to leave engineering studies to specialize in a different field. The student must develop an undergraduate course of study sufficiently broad to serve as a good preparation for entrance into graduate studies in another profession. Early consultation with an academic advisor is a must in this regard. This will save a student much unnecessary time in an undergraduate program, if the studies can be patterned toward this ultimate goal of advanced study in another profession.

GRADUATE WORK IN OTHER DISCIPLINES

In addition to the students mentioned in the previous section who decide to go into another profession rather than engineering and who pursue graduate studies in those other professions, a number of engineering graduates find it helpful to pursue graduate studies in other disciplines but to continue to practice as engineers. In the same way that an engineering education is found to be useful in other professions, a knowledge of law or business may be extremely useful to an engineering consultant or to an engineer who works as an industrial manager. Each year many thousands of graduate engineers enter law schools or enroll in programs leading to Master of Business Administration (MBA) degrees. Sometimes these students find that they must take some additional undergraduate courses which are prerequisites to the graduate level courses in law or business or whatever field of study they wish to enter. These graduate schools in other professions will demand a certain level of excellence in undergraduate studies for admission, so not all graduate engineers will qualify.

CONTINUING EDUCATION IN ENGINEERING

We have referred on previous pages to the explosion of knowledge which has taken place in recent years. The rapid pace of technical change in today's world means that technical practitioners, including engineers, must continue their education or run the risk of becoming obsolete. The amount of technical information available to engineers today is more than double what was available at the end of World War II. The civil engineer, as well as all other engineers, must continue her or his education after graduation, in order to maintain the ability to practice. At graduation, the civil engineer is assumed to be not only knowledgeable about the present state of the art but also capable of expanding and improving her or his body of knowledge by continually keeping informed of all the new developments in technology. This effort to keep up-to-date requires a great amount of time and effort. Many engineers try to accomplish this in an informal way, reading journals and periodicals on their own to stay informed about the latest developments in civil engineering. The many new techniques of analysis, new engineering materials never before available, new applications of computers, and many other new developments can virtually overwhelm an individual. For this reason, many engineers decide to continue their education through short courses or seminars offered by professional societies or universities to large groups of engineers. In some states, engineering practitioners are called upon to demonstrate periodically (every few years) that they are maintaining their technical competency and should be allowed to continue practicing as licensed engineers. In general, the universities and members of the academic community have taken the lead in helping practicing engineers maintain their technical competency. Obviously, the individual engineer has the primary responsibility for maintaining her or his own professional competence and for continuing her or his professional development. However, universities and professional societies also share this responsibility and usually offer a wide variety of programs designed to continue the engineer's education. Some individuals who feel a need for continuing their education in depth will enroll in formal educational programs leading to advanced degrees. The majority of individual engineers, however, usually will try to enroll in short courses or seminars with the specific objective of learning about a particular new development in engineering, rather than trying to obtain a comprehensive background in a new specialization or a new field. A certain amount of continuing education obviously occurs on the job as an engineer works with other professionals or under the direction of another professional. However, because of the rapid pace of technological change and because of the great number of new developments in civil engineering in recent years, most engineers find it necessary to attend short courses and seminars to obtain the required information about new developments.

Many industrial employers will arrange for continuing education programs to be held for large groups of their engineering employees. In many cases, firms pay full tuition or share the tuition cost with the individual engineer who wants to take a single course for one or more semesters, in order to increase capability in a particular area. Many employers also will pay the engineer's costs to go to a convention, short course, or seminar for one or two days. Professional societies often sponsor these meetings, and the individual meetings can vary in length from one day to as long as a week. At these meetings, technical papers are presented by expert practitioners or people who are engaged in research. The idea of the presentations is to bring new information to the practicing engineer in as short a period of time as possible.

An engineer's education will never be complete. A practicing civil engineer will continue to learn throughout her or his entire professional lifetime. Graduate engineers typically will look through professional journals and magazines for notices concerning the time, scheduling, and cost for various continuing education activities such as seminars and short courses. In some cases, it is possible to obtain videotapes or other materials which can be used by the individual at home to increase knowledge about a particular subject. Most of the time, however, continuing education takes the form of attendance at meetings and seminars. Not all of the continuing education efforts which a civil engineer makes will be intended to improve technical competence. Many of the continuing education activities carried out by today's engineers are designed to improve their management skills and their ability to communicate. Many industrial firms conduct special courses in communication skills, interpersonal relationships, and other managerial activities to improve the efficiency and effectiveness of their employees. Engineers often enter the ranks of management in large companies, and consequently, many graduate engineers participate in continuing education efforts designed to improve their management skills.

EDUCATION COSTS

For most students, the costs of a college education will represent one of the major investments in their entire lives. Not only are the four or five years of required education a time lost from gainful employment, to a great extent, but the direct cost of tuition, room and board, books, and other expenses for college life will be very high.

Tuition costs vary greatly, with low costs usually associated with state universities and higher costs associated with private colleges. Some typical tuition costs, based upon the cost for two semesters, are shown below:

Georgetown University	$9,300
Vanderbilt University	$8,500
University of Notre Dame	$7,985
Catholic University of America	$6,650
Wake Forest University	$6,000
University of Louisville	$1,350

Based on 1985 catalogues and brochures distributed by the schools listed.

Costs for room and board vary greatly from one school to the next, but generally will range from about $2,500 to $3,700 per year. Book costs also vary somewhat, but costs for books and supplies will usually be between $350 and $450 per year for most engineering students. Of course, personal expenses and transportation between the university and home will vary from individual to individual, but these are real costs and should be considered by the student in making an estimate of total expense for going to college.

The costs associated with attending a particular university or college can vary widely from the figures given here, and it is important for you to investigate the costs at the particular school you wish to attend. One very important consideration is the differences in tuition rates and other fees for residents of the states in which the schools are located, compared to the fees and tuition costs for non-residents.

Most state-supported schools charge significantly higher tuition rates for non-residents.

FINANCIAL AID OPPORTUNITIES

Since the costs for any college education are very high, and costs for engineering education are somewhat higher than average, it is important for you to investigate all possible sources for financial aid, which can come in many different forms. Basically, the overall process involves the student's making an application, furnishing information about family income and assets as well as financial responsibilities, and the evaluation of the application by the college administration or by a governmental agency responsible for supplying financial aid. Many students may be eligible to receive aid from the Veterans Administration or from the United States Department of Health and Human Resources. Scholarships and loans are available from various organizations and from individual schools and colleges. Many employers as well as many social and professional organizations offer scholarships and loans for students who are in some way connected with members of these groups (family members or future members of particular professions, usually). In some instances, the benefits available to veterans also can be applied to their dependents. Information concerning all the educational benefits available to veterans can be obtained by contacting a local Veterans Administration office. Assistance is available also through the Red Cross for veterans who are unable to make contact with representatives of the Veterans Administration.

The most common source of student financial assistance is the United States Department of Health and Human Resources. This assistance is furnished under a number of different programs. In the Supplemental Educational Opportunity Grant Program, students with exceptional financial needs, who are unable to continue their education without assistance, are eligible for awards of not less than $200 per year and not more than $2,000 per year. To be eligible for this money, the student must be enrolled in an approved institution for at least a half-time program. The awards may be received for up to four years in general educational programs, but in five-year programs such as cooperative work programs or in professional schools with five-year programs, the awards may be extended over the full five-year period.

The Basic Educational Opportunity Grant Program is another source of funds for eligible students. Pell Grants are federal grants awarded to undergraduate students from families with incomes up to $25,000. Full-time students of universities, community colleges, and junior colleges may apply for financial assistance under this program. Usually, applications for this type of assistance are available from high school counselors, from colleges, and from public libraries. Applications may be obtained by writing the Department of Health and Human Services at P.O. Box 84, Washington, D.C., 20044. If the program administrators consider the student eligible for a grant under this program, a notice is sent to the college of the student's choice, and the college administration participates in determining the amount of the actual award to the student. The amount of the award is calculated by the college on the basis of the costs of tuition and the capabilities of the student and her or his family to meet those costs.

Many universities and colleges grant direct academic scholarships on the basis of ranking in high school graduating class and college entrance examination scores. Usually, the required entrance test is either the Scholastic Aptitude Test (SAT) or the American College Test (ACT). Some states have special financial aid grants for residents of those states attending private colleges or universities within the state. Other scholarships are available specifically for women and minority group members, or for students with affiliations to particular groups. Information on this type of financial aid is available from the university directly.

In addition to scholarship awards as described above, financial assistance is also available from the United States Government in the form of student loans. Under the provisions of the National Direct Student Loan Program, students who are enrolled at least half-time in an accredited college and who are unable to fully meet their educational expenses may borrow between $300 and $1,500 per year. The interest rate on these loans is five percent per year. Payments on the principal and interest for the loan must begin six months after the student leaves school. If it is necessary to do so, the student may extend repayment over a long time period at a lower interest charge, upon approval by the funding agency. If the student leaves college to enter the armed forces, the Peace Corps, or Vista, he or she need not repay any part of the loan for the first three years after termination of the course of study. Some loans under this program may require no repayment if the student enters certain professional fields after graduation. Information on the details of this program can be obtained from the financial aid officers in the colleges where the civil engineering programs are offered in which the student is interested. Another program through which a student may obtain a loan is the Guaranteed Student Loan Program. Under provisions of this program, the student may borrow between $500 and $2,500 per year. To be eligible for this program the student's adjusted family income must be less than $30,000 per year, or the adjusted family income can be over $30,000 a year if it is possible for the family to demonstrate important financial need. Under this program the loan is obtained from a private lender such as a bank or credit union. To be eligible for such loans, students must be enrolled for at least a half-time program in an approved college or university. Repayment on the principal and interest of the loan must begin within six months after the student leaves school.

A third loan program is the Federal Parent Loan Program now available in many states. In some states this program is called the Parents' Loan for Undergraduate Students (PLUS). Under this program, it is possible for parents to borrow up to $3,000 at an annual interest rate of 12 percent with up to ten years to repay the loan. Repayment of the loan must begin within 30 to 60 days after the money is received, and the minimum monthly payment is $50. Further information on this program may be obtained by sending a self-addressed stamped envelope and an appropriate request to HEAF, 1030 Fifth Street, NW, Suite 1050, Washington, D.C., 20005.

Civil engineers are best known for designing and building bridges, such as this modern structure spanning a canal in Tennessee. (American Society of Civil Engineers photo)

Work-Study Programs

In addition to the scholarships, grants, and loans described above, a student may obtain financial assistance through work-study programs financed by the university itself or by the federal government. The College Work Study Program is funded by the Department of Health and Human Services for the purpose of providing jobs for students who have extreme financial needs. To be eligible for this program, the student must be enrolled at least half-time in an approved institution. The particular job assignment for the student is arranged by the college and may be in campus laboratories, offices, libraries, or cafeterias. The grant under this program varies from $1,000 to $1,500 per year. In some cases, a university may place a student in a position off campus with a public or private non-profit organization such as another school or a hospital. Students may be employed for as many as 40 hours per week and must receive salaries equivalent to the current minimum hourly wage. Information and applications for this assistance can be obtained from the financial officers of the individual schools.

Applying for Financial Aid

In trying to obtain financial assistance, the student and the student's parents must supply a considerable amount of information about their financial capabilities and needs. This information is usually supplied on a standard form. The two forms which are most commonly required are the Financial Aid Form (FAF) of the College Scholarship Service (CSS) and the Family Financial Statement (FFS) of American College Testing, Inc. In most cases, parents should submit the FAF to CSS providing actual income data no later than January 15. Some universities and colleges require students to submit the most recent year's federal income tax return for their family. The requirements for information will vary somewhat from one school to another and it is important for you to obtain exact requirements from the schools in which you are interested. It is important to remember that there are many sources of financial assistance available to prospective engineering students. Most colleges and universities have special scholarship programs and loan funds for students in selected programs. Many sources of financial aid often are overlooked. These sources include parents' employers, local clubs, fraternal organizations, veterans associations, youth groups, religious organizations, chambers of commerce, and other societies and organizations. Every effort should be made to investigate these possible sources of assistance. A student should seek help from a school counselor and from the financial aid officers at the schools he or she might like to attend. Many times, students fail to receive scholarships or other forms of financial aid simply because they are unaware of the possibilities for such aid or because they apply too late.

Almost every university has a special financial aid office with responsibility for administering aid programs at that particular school. Usually, it is a good idea for a student to obtain all the available information from the financial aid office as soon as the student becomes interested in that particular university. Then the student and his or her family can submit the parents' confidential statement, the statement of the student's own financial situation, or any other required information to the Financial Aid Office. All of these forms should be submitted at the earliest possible date in order to make sure that the student has an ample opportunity to secure financial aid.

Surveying is one of the many areas in which civil engineers may specialize. (photo by John H. Woodburn)

CHAPTER 3

SPECIALTIES WITHIN CIVIL ENGINEERING

In the first part of this book, the history and scope of the field of civil engineering were discussed. That discussion was intended to give you an overview of activities included in civil engineering; however, it is not possible to describe all of the activities of civil engineers in a book such as this. It also is difficult to include a description of all the ways in which civil engineers work with other engineers, scientists, and non-technical personnel. What we hope to accomplish in this chapter, instead, is a description of the major technical specialties and some of the functional activities within the broad field of civil engineering.

For our purposes in this book, it is possible to group all of the activities of civil engineers into various technical categories, such as structural engineering, transportation engineering, environmental engineering, etc. The activities of civil engineers working in any one of these technical specialties also can be divided into several functions, such as research and development, planning and design, supervision of construction, and operations. In addition, sales and management are viable activities for civil engineering graduates. The technical fields of specialization will be described first, followed by the different functional activities of civil engineers.

63

TECHNICAL SPECIALTIES

The American Society of Civil Engineers has established a number of technical divisions within its organization, and these divisions give good indication of the various technical specialties within the general field of civil engineering. The ASCE divisions include:

Surveying and Mapping
Water Resources Planning and Development
Aerospace
Hydraulics
Irrigation and Drainage
Waterways, Ports, and Harbors
Structural
Engineering Mechanics
Highway
Air Transport
Pipeline
Urban Planning and Development
Urban Transportation
Construction
Geotechnical Engineering
Environmental Engineering
Materials Engineering
Energy

ASCE also has technical councils on other specialty areas including aerospace engineering, computer practices in civil engineering, earthquake engineering, cold regions engineering, and ocean engineering and research. The work carried out by civil engineers in these various specialty groups can best be described by combining those activities which are most often closely associated in practice. Such combination will be used in the descriptions that follow.

SURVEYING AND MAPPING

The field of surveying and mapping is as ancient as historical records. Land surveys were conducted in ancient Babylonia and

Egypt, more than 4,000 years ago. Surveying and mapping efforts also played a very important role in the development of the United States. These surveying expeditions included that of Lewis and Clark, who traveled up the Missouri River and then on to the Pacific Ocean in 1803, and the expeditions which led to the development of the American West in the 1850s and 1860s. A major new development in surveying and mapping occurred in the 1920s when airborne cameras were used to produce photographic maps. This type of mapping now has been carried to even greater sophistication through the use of the mapping satellites which now circle the globe.

One of the activities included in the general field of surveying and mapping is called cartography. Cartography includes all those activities in the field necessary to the development of maps, as well as those activities in the office associated with drawing, editing, and printing of the maps. Cartographers prepare maps and charts from data gathered by physical surveys, by aerial photography, and by the more sophisticated means of exploration, such as remote sensing using various forms of radar and other techniques. The products of cartographers include topographic maps showing the elevation and major physical features of the land surface; nautical and aeronautical charts; meteorological maps; city maps; political maps of countries; atlases; and others. Clearly, the most critical element in producing maps is the preservation of accuracy and detail throughout the entire process. Precise surveying techniques are utilized, and precision drafting instruments, aerial cameras, computers, printers, and plotters are used. In order to convert numerical and physical data into maps to be used by pilots, drivers, and pedestrians, the civil engineer practicing cartography calls upon many aspects of her or his engineering education and experience and, particularly, knowledge of geography and skill as a designer and artist. In order to produce maps of various kinds, this civil engineer depends upon information supplied by surveyors.

Surveying is principally concerned with the determination of boundaries of real property, and with the precise location of route alignments. Surveying also includes the planning and subdivision of areas and the preparation of deed descriptions for conveying ownership of parcels of land. Obviously, the civil engineer involved in surveying must work closely with draftspersons and cartographers in the preparation of maps showing natural and human-made features on the land, the land and water surfaces in a given area, and the boundaries of pieces of property. The work of the surveyor includes the establishment of state and local boundaries for the definition of townships, ranges, and sections. In addition, today's surveyors are heavily involved with computer applications. Much of the field equipment, in fact, is computer-driven or uses computers in field data acquisition and storage. There is also a natural link between the surveyor and CAE/CAD (Computer-Aided Engineering, Design and Drafting), in that quite often surveying data gathered onto disks or tapes in the field is processed into maps and charts by computer without anyone's ever putting pencil to paper.

In most cases, surveyors in private practice are registered/licensed in one or more states. Today, in the United States, there are approximately 40,000 licensed surveyors. The techniques utilized by the civil engineer in surveying are quite similar to those used in other engineering surveys.

Engineering surveys include those measurements which provide the data utilized by engineers in planning, locating, designing, and constructing physical facilities. Among the basic data used in design are the horizontal and vertical locations of natural features of the land surface, the location of man-made features, and the depiction of those conditions through the development of cross-section and profile views showing the elevation and character of the ground surface along lines established for design and construction purposes. Civil engineers involved in surveying are responsible for the layout of structures to be built, the measurement of quantities of materials used in construction, and often the actual performance of surveys in mines, quarries, and other excavation activities.

In recent years, civil engineers involved in surveying have become increasingly dependent on aerial photography as an important aid in surveying and mapping. In such work, cameras are installed in specially designed aircraft which fly at predetermined altitudes along straight lines across the land area to be mapped. The flight paths of the aircraft are arranged so that the land areas photographed each time the camera shutter is opened will overlap. The product of the aerial survey is a series of carefully calibrated photographs which can be placed in a defined sequence along lines which correspond to the flight paths. The height of the aircraft and the camera angle are adjusted so that the overlap of the photographs permits the cartographer to use optical devices to produce a three-dimensional view of the land surface. Mechanical devices and instruments have been devised to use with aerial photographs so that topographic maps showing the elevation of the ground surface at any particular point can be prepared mechanically. This type of surveying has been greatly refined and now can be applied to the entire surface of the globe through the installation of cameras in satellites. Recent developments in three-dimensional computer modeling have given engineers the ability to generate terrain models of the land by converting coordinate data graphically.

With such optical devices as transits, levels, and theodolites, as well as extensive reliance on computers, civil engineers now are able to prepare maps of the surface of the earth, charts of the ocean bottom, boundary lines to private and public lands, etc. This branch of civil engineering is among the oldest of the specialty areas within civil engineering, yet is becoming one of the most dependent on high technology. At this time, about one in 20 of the total number of civil engineers in the United States is involved in the practice of surveying and mapping.

WATER RESOURCES ENGINEERING

Water resources engineering includes the activities of civil engineers who are engaged in obtaining adequate supplies of drinking water and water for industrial uses; engineers who design and construct structures to retain and transport water; engineers who are concerned with the relationship between rainfall and the amount of water stored in the ground as well as that which flows across the land surface; and engineers engaged in irrigation and drainage of agricultural lands. The general area of water resources engineering includes many civil engineers who are primarily concerned with maintaining water quality. These engineers work to purify drinking water supplies and to treat wastewaters so that they can be released into streams and other bodies of water. The activities of this last group of engineers will be described in a later section on environmental engineering.

Among the most important activities of water resources engineers is the development and conservation of adequate supplies of water for drinking water needs and for use in industry. Water resources engineers also are engaged in supplying water for irrigation of farmland, particularly in the arid areas of the American Southwest. In certain regions of Texas, Oklahoma, New Mexico, and Arizona, ground water depletion and the resultant lowering of water tables has produced urgent problems that must be solved in order to maintain economic viability. Civil engineers who specialize in water resources are working in cooperation with geologists to develop technically feasible and cost-effective solutions to these water problems.

Supplying adequate quantities of water for today's industrialized society in America is a tremendous undertaking. Drinking water supplies are a major item; the average American uses between 150 and 200 gallons of water a day for drinking water, bathing, washing automobiles, watering lawns, washing clothes, and for many other domestic uses. Not surprisingly the water requirements of industry are much greater than the demands for domestic water supplies. For example, over 100,000 gallons of water are utilized in the manufacture and assembly of an automobile, and more than 80,000 gallons of water are utilized in making a ton of steel. On the other hand, the average water requirement for irrigation in the United States is more than one million gallons of water each year per acre of irrigated farmland. The challenge of ensuring reliable supplies of water for all these uses falls to civil engineers who are engaged in water resources engineering.

Water resources engineers often devote much of their time to activities in hydraulics or hydrology. Hydraulics is the study of the movement of water in response to applied forces and of the forces and pressures which are created by both moving and stationary bodies of water. Hydrology, on the other hand, is the study of the occurrence of water in the atmosphere—the fall of water to earth as precipitation in various forms, the flow of water across the surface of the earth, and the infiltration of water through the surface of the earth to underground storage areas. Civil engineers involved with hydrology work to determine the availability of water supplies for various uses; in other words, the attempt to determine how much water can be used safely for particular purposes without permanently depleting and exhausting the supply of available water.

In controlling the flow of water across the land surface and in storing surface water and groundwater in various locations, water resources engineers often rely on dams, floodwalls, pumping stations, aqueducts, and canals. Dams are constructed to store water for irrigation supplies, for municipal supplies, and for the generation of electric power. In almost all cases, dams, pumping plants, and associated pipelines or aqueducts are designed to satisfy many separate purposes so that the most efficient overall use of the available water is achieved. The designer of such water-retaining structures as dams and floodwalls works with irrigation and drainage specialists in designing and constructing canals and other facilities for carrying the water to areas where rainfall is deficient. More than 35 million acres of land are irrigated at the present time in the United States, principally in the Southwest. The work of irrigation and drainage engineers in the United States today is akin to the work of the ancient Egyptian engineers mentioned in Chapter 1, who worked to distribute the flood water of the Nile across their valley homeland. It is too bad that the Egyptian "civil engineers" did not have access to the computers and satellites that their American successors have.

Some water resources engineers are also concerned with the planning, design, and construction of facilities to provide navigation on waterways throughout the United States. In addition, other water resources engineers design and construct floodwalls and levees to hold back maximum flows in rivers to prevent flooding. Flood control also can be achieved through the construction of reservoirs which retain rapid runoff and maximum stream flows, so that these quantities of water can be released gradually to prevent flooding downstream.

Another exciting area also included in water resources engineering is the planning and construction of harbor facilities along the nation's coastlines. Harbor engineering includes design and construction of artificial channels, canals, piers, and docks; dredging of navigation channels in harbors; and the protection of shorelines from erosion by wind and wave action.

All of these activities are included within the area of civil engineering which is called water resources engineering. About one of every five civil engineers in the United States is engaged in water resources engineering.

STRUCTURAL ENGINEERING

The area within civil engineering that produces some of the most dramatic products is structural engineering. Civil engineers involved in this area are concerned with the planning, design, and construction of all types of physical facilities, especially bridges and buildings. Structural engineers do not actually build the bridges, skyscrapers, and factories, but they are responsible for their planning and design. In many cases, the structural engineer does have construction inspection and supervision on the structure he or she designs.

Structural engineers are concerned with the property of such materials as steel and concrete, and they are very interested in new developments in the production of these materials. For example, steel alloys are now being produced which provide much greater strength than was formerly thought attainable. These new, high-strength materials allow structural engineers to conquer space with beams, girders, and columns in an ever more efficient fashion, and help conserve valuable raw materials for future generations.

Some structural engineers are involved principally with the behavior of materials subjected to either static or dynamic loadings. These civil engineers often are found in universities and in research institutions, and they are likely to be members of the Engineering Mechanics Division of the American Society of Civil Engineers. The structural engineer involved with design, on the other hand, is concerned with using the materials developed by materials scientists and manufacturers to design and plan factories, warehouses, commercial buildings, power plants, and all of the other structures which can be found in the modern city. The principal problem faced by structural designers is the use of available materials in the most efficient (and cost-effective) way to satisfy the desires of clients, usually the owners of the proposed structure. Obviously, this civil engineer must also adjust his design to the conditions on a particular selected site. The finished structure must be capable of performing its desired function or purpose, whether it is located on a hillside in Oregon or in a cypress swamp in Florida; in rain, wind, snow, and all of the other extremes of weather to which a structure may be exposed; or in a zone of potential earthquake (or seismic) activity.

One of the most important fields of activity for today's structural engineers is the planning and design of power plants and their associated structures, especially nuclear reactor containment facilities. The containment vessels used to house nuclear reactors obviously must satisfy very strict safety requirements for the protection of the public. The structural engineer also must be extremely careful in designing the foundations in power plants where coal is burned to produce steam, because steam-powered turbine-generator units for the production of electricity are very sensitive to settlement.

Perhaps the most dramatic structures designed by civil engineers are bridges. Bridges come in a variety of forms and shapes including arch bridges, cantilever bridges, truss bridges, and—most dramatic of all—suspension bridges. Outstanding examples of suspension bridges include the Golden Gate Bridge in San Francisco and the Verrazano Narrows Bridge between Staten Island and Brooklyn, in New York. These graceful structures, spanning thousands of feet from shoreline to shoreline, seem to leap from one bank to the other and are among the most artistic products of civil engineering. Most importantly, these beautiful structures, as well as all other types of bridges, must be designed to carry heavy traffic loads in all types of weather, and must be capable of withstanding any natural forces produced by wind, tide, and earthquake. Civil engineers involved in bridge design are faced with the challenge of developing a structure at the lowest possible cost, without compromising safety or maintenance provisions, or reducing the capacity of the structure in any way. To meet this challenge, structural engineers have access to new materials of greater strength, more durability, and lower weight, developed by other engineers and scientists engaged in research and development activities. The work of the modern structural engineer requires a comprehensive knowledge of the properties of materials as well as a deep appreciation for the balancing of forces in complex, three-dimensional structural frameworks. These skills are perfectly adapted to the use of computers, and as a result, the practice of structural engineering is highly dependent on expertise in computing.

Structural engineers are the most numerous civil engineers in the United States today. Approximately one out of four civil engineers in the U.S. is involved in this exciting area.

TRANSPORTATION ENGINEERING

Civil engineers engaged in transportation engineering are concerned with all the modes of transportation, including rail, transit, highways, aviation, areospace, and pipelines. Whatever the mode of concern, the transportation engineer of today is vitally linked to and dependent upon computers for such activities as data acquisition and management, model execution and "number crunching," graphics, and so on.

The railroad engineer of the past has given way in many localities to the highway engineer, since highways have now replaced railroads as the major carrier of people and, to a great extent, freight in the United States today. Nevertheless, more than 300,000 miles of railroad track still remain in service in the United States. In order to meet the increasing costs of maintenance of track and roadbeds, civil engineers are often engaged in rebuilding old railroad lines and improving the combined structure of the rail, the tie, and the supporting ballast beneath the track. Occasionally, new lines must be built to serve new industrial and commercial facilities; in fact, in recent years, a combination of truck freight with railroad transportation (the "piggy back" service in which truck trailers are carried long distances on railroad flat cars) has developed to respond to industrial and commercial opportunities. This, in turn, has led to the demand for new terminal facilities designed to accommodate the movement of a large portion of the nation's raw materials. Surprising, many of these raw materials, including ores and fuels such as coal, still travel by rail, as do many types of manufactured goods (heavy industrial machinery, new automobiles, heavy military equipment, etc.).

In urban areas, railroad rights of way are being redesigned to satisfy mass transportation needs. Urban mass transportation elements include the planning and design of subways, elevated rail lines, and surface rapid transit routes. Rail/mass transportation engineers' responsibilities in accomplishing all of these activities are in some cases similar to the work carried out by highway engineers.

Highway engineers are responsible for the planning, design, and maintenance of highway systems and secondary roadways. The work of highway engineers begins with the planning of highway facilities to meet future needs created by changing population patterns and the growth of cities and towns. Their work, in this respect, is similar to that which was done by the railroad engineers during the development of rail systems in the United States more than 100 years ago.

Highway engineers must analyze basic traffic patterns and attempt to predict future traffic flows. After determining the need for highway facilities on the basis of traffic predictions, they must develop the most economical combination of highway alignment and location, while paying much attention to environmental and social impacts. Highway engineers are involved in the construction of embankments and fills to carry roadways across low areas and of cuts through hills and mountain ranges. Highway pavements must be designed to reflect most efficient use of asphalt, concrete, and other materials, available at different prices in different locations. All of these factors—available materials, alternate routes, environmental and social impacts, traffic demands—must be considered by the highway engineer in developing the most efficient transportation system for the lowest expenditure of money, materials, and labor.

A more specialized area of transportation engineering activity is the planning and design of airports and their related facilities. A futuristic activity of transportation engineers, in this regard, is the planning and design of space ports and other facilities associated with the exploration of space. However, such glamorous activities occupy the time of only a very few civil engineers. Many more are concerned with the design and construction of airport facilities, including hangars, runways, taxiways, parking aprons, and terminal facilities. These facilities must be adequate not only for the landing, movement, and takeoff of aircraft, but also for the efficient movement of passengers, freight, and the equipment necessary for aircraft maintenance and service. Certainly one of the most important and controversial elements of airport and aviation engineering is the mitigation of environmental impacts, particularly noise. The Federal Aviation Administration has invested literally billions of dollars in noise mitigation since the 1960s.

A less obvious form of transportation engineering is the planning and design of pipelines to carry liquids, gases, and combinations of liquids and solids. The pipeline project most familiar to readers today probably would be the design and construction of the "Alaskan Pipeline" to bring petroleum from the North Slope of Alaska to consumers in the "lower 48" United States. However, many other pipeline projects of great importance have been planned and designed by civil engineers specializing in this area. Natural gas pipelines were constructed throughout the United States in the years following World War II to provide Americans with one of the most convenient forms of energy for use in habitat heating and industrial processing. It is also possible to move slurries (mixtures of solids and water) efficiently over long distances in pipelines. Pipeline transport of coal slurries is assuming increasing importance today as we are faced with possible shortages of natural gas and petroleum. Civil engineers who specialize in pipeline transport of materials must determine the most economical alignment of pipelines, with full consideration for the possible social, political, and environmental effects of such construction. These engineers plan and design power plants, terminal facilities, receiving and discharge plants, and pumping stations associated with pipelines. Transportation engineers engaged with pipelines work closely with structural engineers and with geotechnical engineers and other specialists in the broad field of civil engineering.

As you can see, transportation engineers have varied opportunities for involvement. Approximately one out of seven civil engineers in the United States is in the transportation engineering area.

PUBLIC WORKS/URBAN PLANNING

More than three-fourths of the population of the United States lives in cities. Consequently, much of the work of the civil engineer today is associated with the design and planning of structures and facilities to serve urban communities. Civil engineers who are involved in urban planning and development work with other planners and designers in associated fields to accomplish a variety of tasks. Among other interesting duties, these engineers may predict, or "model," the future needs of a community with respect to streets and roadways, recreational and park facilities, freeway and airport locations, urban renewal projects, industrial parks and associated developments, and residential areas. Civil engineers engaged in urban planning may serve as members of planning commissions or zoning boards which act in an advisory capacity to elected municipal officials. In many cases, the civil engineers' contributions as members of zoning commissions will be to provide the necessary technical expertise required for decisions on how the control of community growth is balanced and coordinated. The engineers will work closely in these activities with planners, architects, politicians, citizen groups, and other persons concerned with the prudent and careful planning of future urban development.

The day-to-day maintenance activities of cities and towns are the responsibility of public works engineers, who generally are civil engineers employed by municipal or county government. These engineers often act in a supervisory capacity and coordinate the efforts of consultants and other civil engineers who are working on the design and construction of specific projects. Public works engineers are responsible for the coordination of activities associated with water supply systems; drinking water distribution lines; wastewater treatment disposal systems; municipal streets, roadways, and parking facilities; solid waste collection and disposal systems; schools,

hospitals, libraries, and other types of public buildings; and munici-
pally owned service facilities such as power utilities. Public works
engineers must be educated in a broad spectrum of civil engineering
activities and must be capable of managing and coordinating the efforts of many other engineers and non-technical personnel.

Because so much investment must be made to maintain our nation's urban infrastructure (roads, streets, bridges, sewers, etc.), the
future is particularly bright for civil engineers who desire a career in
public works. Currently, about one in eight civil engineers is working
in this area.

CONSTRUCTION ENGINEERING

Construction engineering is one of the newer areas within civil engineering, even though construction supervision is quite a traditional engineering activity. Construction engineers are educated in the
general field of civil engineering, specializing in the management
and control of construction activities. These engineers are vitally
concerned with the planning and scheduling of every phase of construction activity on a particular project, from the initial order of
materials from a supplier, through to the time when the owner takes
possession of the finished structure. Construction engineers must
plan the entire project, including the selection of the equipment, materials, and labor that will be needed at any particular location on any
given day. Thus, these engineers are heavily engaged in scheduling
and programming activities. In most cases, construction engineers
are employees of construction companies, or they are the owners of
such companies. Recent civil engineering graduates opting for careers in this field are likely to be involved with testing the materials
of construction and in inspecting the project site to make certain that
it complies with the designers' specifications. As construction engineers gain more experience, they are given responsibility for
supervising construction activities; with continued experience, they
are likely to assume management roles in construction companies.
Experienced construction engineers will be responsible for hiring
site personnel and construction specialists, for estimating costs

based upon structural designs, for monitoring day-to-day expenses on the job site, and in the competitive process of bid preparation. Construction engineers must work closely with designers and with the owners of the structures being built to ensure that the owners' wishes and desires are carried out in the most efficient and cost-effective manner.

The earning potential for civil engineers in construction is the highest for all the areas in civil engineering. However, the demand for these engineers is quite sensitive to the existing state of the economy. While the number fluctuates, we estimate that about one in ten civil engineers practicing is in the construction area.

GEOTECHNICAL ENGINEERING

Another rapidly growing specialty group within the field of civil engineering is geotechnical engineering. Geotechnical engineers are responsible for determining the characteristics of the soils and rocks at a given construction site or along the route of a transportation facility. Geotechnical engineers apply the principles of mechanics to the natural soil and rock materials that are found on project sites. They develop designs for foundations to support bridges, industrial buildings, and other structures. They also develop excavation techniques and construction methods for tunnels and other works.

Geotechnical engineers are responsible for conducting exploration and testing of the geological materials which will be the foundations of the structures included in particular projects or which will form the construction materials for embankments for highways and railroads or for dams, levees, and floodwalls. In so doing, these engineers work closely with engineering geologists to detect possible weaknesses in the soils and rocks at particular sites and with structural engineers and designers to match the characteristics of the soils and rocks found at the site of the proposed facility. Today, about one in ten practicing civil engineers is working in the geotechnical area.

ENVIRONMENTAL ENGINEERING

A relatively new area in civil engineering has matured in recent years—environmental engineering. For many years, civil engineers were involved in obtaining and purifying drinking water supplies and in designing and planning facilities for the removal and treatment of wastewaters from residences and industrial facilities. Civil engineers specializing in this activity were traditionally referred to as "Sanitary Engineers." In recent years, the role of pollution control has been greatly expanded to include efforts directed toward air pollution control, solid waste management, and environmental impact analysis, as well as the more traditional activities of water pollution control. This broadening of interests has resulted in the common use of the designation "Environmental Engineer."

Civil engineers engaged in this area work closely with ecologists, biologists, and other scientists to determine the quality of life in particular environments. They try to develop engineering systems to accomplish particular goals with minimum damage to the natural environment. These pollution control activities include the design and planning of wastewater treatment facilities, flood control facilities, groundwater recharge areas, air pollution control facilities, and resource recovery plants.

This specialty within the broad field of civil engineering expanded rapidly during the 1970s and has now reached a stable level of activity. Today, about one in eight practicing civil engineers works in environmental engineering.

(*Note to the reader:* If you were keeping track of the number of civil engineers working in each area out of the total number of practicing civil engineers, you would have noticed that we accounted for 107 percent of the total. This is possible because many civil engineers are specialized in more than one area.)

OTHER SPECIALTIES

A number of other specialty areas of engineering attract limited numbers of civil engineers. Included in these are design, construction, and operation of power plant and transmission facilities; computer-aided engineering (primarily software development applied to civil engineering problems); applications of systems science (optimization techniques) to the planning and design of public works facilities; design and construction of ocean-floor and offshore structures, particularly structures associated with petroleum exploration, removal, and transport; and so on. Civil engineers find little difficulty in utilizing their basic education in the sciences and in accumulated civil engineering experience to be successful in these areas of activity. Within each, civil engineers work closely with a number of engineers and scientists in closely affiliated specialties.

FUNCTIONAL SPECIALTIES

Many civil engineers are engaged in planning and design activities, and great numbers are involved in supervision of construction and the operation of engineered facilities. In addition, a significant number of civil engineers also are engaged in research and development activities. Others who have been educated and received their primary experience in civil engineering have moved into sales of equipment and systems designed to accomplish many of the purposes of these engineered facilities. Engineering sales requires certain characteristics, not the least of which is a very comprehensive and adequate background in engineering. Finally, some civil engineers are engaged in managing the activities of their fellow engineers; some reach executive levels of management where they supervise and coordinate the activities of scientists, engineers, technicians, craftsmen, and other non-technical employees in large corporations.

RESEARCH AND DEVELOPMENT

In the field of civil engineering, research and development activities are largely carried out by university professors and engineers employed in research institutes. In addition, a significant amount of civil engineering research is carried out in the laboratories and research organizations of federal and state governments. Federal and state governments support about one-half of all the civil engineering research activities carried out in the United States today. However, it is not likely that this government support will increase, because of budget constraints at the federal level. On the other hand, our nation needs the results of civil engineering research now more than ever because of the ever-worsening condition of the infrastructure. To help meet this need, the National Science Foundation (the most prestigious research organization in the United States) is focusing increased attention on engineering research on infrastructure issues.

In most research and development organizations, a specific problem or need is identified and then a team of engineers and scientists is established to find a solution to the problem. In most cases, the team will include scientists who are knowledgeable in basic principles and experiences in carrying out laboratory experimentation. The engineering members of these teams are more likely to be involved in the application of the scientists' findings in a practical way to the solution of the problem at hand. Research engineers must perform a coordinating function between the work of the basic scientists on the research team and the engineering designers or the consumers who have identified the particular problem under investigation or who have requested a particular project.

The individual abilities and attributes necessary for successful work in research and development include high intellectual capabilities, exceptionally strong mathematical and computational skills, and a well-developed imagination. Research team members must have a lot of patience and self-confidence in order to continue their investigations, even if many of their efforts prove unsuccessful. It is necessary for research and development team members to be open to new ideas. They must be able to work well with others, sometimes in situations where everyone on the research and development team feels a great sense of frustration with their continued failure to find a solution to the problem at hand.

Engineers working in research and development also must possess some other special qualifications. They must be adept at making accurate observations from experimental data and interpreting those data to draw the correct conclusions. After coming to a definite conclusion about a particular problem and developing a solution to that problem, the engineers must be able to convince the other members of the research and development team, and more importantly, the other members of the affiliated professional organization (such as ASCE), of the quality and correctness of their solution. Research and development engineers must be able to present their ideas convincingly to their immediate supervisors and to top-level management. However, they must be willing to change their conclusions if the evidence presented shows that their solutions are unsatisfactory in terms of cost or convenience to the user or client. After research and development engineers have determined a workable solution to a given problem, they usually transfer their ideas and conclusions to another team of engineers who apply the results of their research efforts to the solution of specific problems. These planners and designers must take the basic concepts of the solution developed by the research and development team and turn those concepts into a set of definite plans and schedules for the construction and operation of a particular facility.

PLANNING AND DESIGN

One of the most important functions of civil engineers involved in planning and design is gathering basic data on the specific location at which a facility is to be constructed and on the specific needs and services at that specific location. In other words, in taking the results of research and development activity and applying those results in actual practice, civil engineers must adapt the results of research and development to actual field conditions. To do this, they must obtain comprehensive data on the site conditions; for example, they must test the soils and rocks at the project site; develop topographic and geologic maps; test environmental conditions in the area (air and water quality and water supply); and determine the technical characteristics of the human-made facilities in the immediate area of the proposed project. The investigation of these facilities may include determining the location, depth, and character of foundations of existing buildings which cannot be disturbed by the construction of a new, adjacent structure. On the other hand, investigation of facilities may include accurately locating all of the important structures in a particular area so that the route for a rapid transit system or highway may be selected in the most efficient and least damaging location.

Obtaining the basic data necessary for the planning process may consist of determining the number and characteristics of the population in the area to be serviced by a particular project and of predicting the changes in number and character of the population in that area in years to come. After obtaining basic data of this type, civil engineers develop both long-range and short-term plans to meet the needs of the community.

In preparing a design, civil engineers must take abstract concepts and transform them into a set of plans, drawings, and accompanying specifications. These in turn are used by contractors and builders as they transform concepts and ideas into a physical reality in concrete, timber, steel, and other materials. Design is the function which is intermediate between planning and actual construction. Specialists in other branches of engineering, such as mechanical and chemical engineering, often are engaged in formulating designs to be used in manufacturing processes. In contrast, the design civil engineer almost always is involved in developing a design package for construction purposes.

When formulating a design, civil engineers seek to make the best fit between the basic concept of a bridge, a structure, or a tunnel and the conditions existing at the site of the proposed facility. They select materials and construction methods with consideration for accomplishing the purposes of the basic design but also with the knowledge of the availability and costs of materials at the site of the proposed facility. Materials for construction and methods of construction must be selected with an appreciation for the limitations of the labor force available in a particular location. For example, civil engineers who are designing a bridge to be built in a remote jungle location must recognize that both prefabricated materials and highly skilled laborers will have to be transported to the construction site. If it is possible to utilize locally available materials and unskilled labor in the construction of parts of the bridge, it may be possible to save considerable sums of money. In this case, the most sophisticated design for the bridge may not be the best design with respect to the overall cost of the structure. In this particular example, the research work of materials scientists in developing high-strength steel alloys will be of little value if designers conclude that transporting such materials to the remote location would be more expensive than using greater amounts of locally available materials.

Additionally, design civil engineers must be capable of preparing plans and specifications which are easily understood by the individuals who are responsible for the actual construction. They must develop specifications which are clear and completely understandable by the laborers on the project site. Civil engineering designers must always be conscious that the lives and fortunes of many persons will rest upon the adequacy of their designs and upon the clarity and quality of the plans and specifications they prepare.

Civil engineers obviously must be creative in order to develop designs which satisfy conditions at specific locations. Additionally, they must be able to imagine solutions which do not yet exist physically. These solutions will not exist until the civil engineer has formulated appropriate designs, and structures have been built according to those designs. Designers must be able to express themselves clearly and accurately in both plan preparation and written instructions for the construction personnel. They must be able to choose between alternative methods of accomplishing the same purpose on the basis of engineering feasibility, cost, and environmental effects.

In almost all cases, civil engineers will be called upon to act as part of design teams. The days of the single designer for a major structure have long since passed. Designers must be capable of working well with the other members of the team, with the members of the research and development group to whom they turn for new ideas, and with the members of the field construction group responsible for bringing their designs into reality.

CONSTRUCTION SUPERVISION

Civil engineers who are primarily interested in supervising and managing construction activities in the field must be capable of taking plans and specifications from design engineers and transforming them into a finished product. This is by definition a difficult task since the plans for major structures often consist of hundreds of pages of drawings, varying from simple overall sketches of a proposed facility to complex diagrams of structural details. They also may include hundreds of pages of written specifications describing the methods to be used in accomplishing the proposed construction activities and setting strict limitations on procedures. Most civil engineers engaged in such activities are involved in the construction of bridges, highways, structures, and water and wastewater treatment-transport systems. In working on these projects, civil engineers will be called upon not only to supervise actual field construction activities but also to estimate project costs and to provide total project management.

As mentioned earlier, engineers are responsible for accomplishing particular tasks with a minimum expenditure of money, materials, and labor. This is especially important to the work of construction engineers, who must develop a mental picture of the construction project from start to finish and schedule the arrival of materials, the use of equipment, and the employment of large numbers of specialists and laborers.

Naturally, construction engineers attempt to select the method of construction which will be most economical. Certainly, they must minimize the cost of construction activities, but before they can do this, they must be able to make preliminary estimates of those costs in order to identify excessive expenditures. Engineers working for owners or developers of proposed facilities also must be able to utilize construction estimates to evaluate bids of contractors wishing to do the work. Conversely, engineers working for bidding contractors must have the ability to make accurate cost estimates for the preparation of bid prices on proposed facilities. The preparation of bid prices is an important task of construction engineers. If their estimates are too low and contracts are awarded on the basis of their low bids, the contractors for whom they work may suffer financial loss when construction costs exceed their estimates even slightly. On the other hand, if a construction engineer makes too high an estimate, the contractor is likely not to be awarded the contract. Obviously, construction engineers working for contractors must develop a fine sense of balance in assessing costs and in judging the possible problems which may arise on a particular construction project.

Once a contract is awarded to her or his company, the construction engineer is responsible for supervising and managing the actual field work. The first step in the process is the actual layout of the project on the site by an engineering survey party. The survey party will place stakes and markers to show the construction crew where to remove soil and rock in order to establish the proper base elevations for the proposed structure. After the surveying crew has established the proper grades and locations, the construction engineer must supervise the building of access roads to the project site. Depending on the type of construction activity, structures may need to be built to house the labor force and to shelter the equipment and supplies used on the project. The construction engineer must be in constant communication with the design engineer so that, if the conditions on the site encountered during construction appear to be different from those assumed before construction began, the construction engineer can so inform the designer, who in turn will modify his or her plans and specifications to meet the changed conditions. This close communication between field construction engineers and office designers is extremely important.

Another of the prime responsibilities of the construction engineer is the supervision and management of the labor force at the construction site. A civil engineer who wishes to be successful as a construction supervisor must be capable of maintaining good relations with the individual workers on the site and with the group of workers as a whole. He or she must be constantly aware of the worker's needs and certainly must be aware of their safety at all times.

Obviously, the construction engineer's work is extremely important to the successful implementation of the designer's plans and specifications. While not all civil engineers are cut out to be responsible for construction activity, those who are find it to be one of the most rewarding areas of practice.

ENGINEERING SALES

Some individuals obtain their basic education in the field of civil engineering and then elect to represent various manufacturers in the sale of equipment and materials utilized in construction activities. To be successful, the sales engineer must possess a fundamental education in engineering, because it is not very productive to try to sell materials and equipment to architects, engineers, and owners of complex and expensive facilities by means of a smooth line of sales talk and a collection of old jokes. The successful construction equipment and materials salesperson should possess a strong background in engineering, preferably in civil engineering.

Obviously, the sales engineer must become intimately familiar with the machinery, products, or materials he or she is trying to sell. He or she must be capable of operating or utilizing the company's products in an expert fashion and must be able to introduce new ideas for their use to potential customers. He or she also must be very skillful in understanding and responding to their complaints about furnished materials and machinery. He or she should be capable of describing to her or his employer's design engineers the needs and desires of those customers and of identifying for the designers the major problems in the field use of the products. Also, he or she must be aware of competitors' developments in machinery and materials.

Most graduates from civil engineering curricula pursue careers in more traditional engineering rather than in sales, but the importance of a background in civil engineering cannot be overemphasized. For those civil engineers whose interests and abilities are suited to sales careers, the potential for financial reward is great.

MANAGEMENT

It is difficult to precisely define management in terms of an engineering career, because almost all engineers are engaged in some aspect of management. More specifically, engineers will be responsible for supervising their fellow engineers in a design office or on a research team, or managing a construction project in the field, even though they may not consider themselves to be a part of the "management" effort.

Management generally is characterized by decision making. The basic duty of the manager is to choose between alternative ways to utilize people, equipment, and materials in order to accomplish a particular task. From this viewpoint, any type of supervisory or decision-making activity on the part of engineers could be considered to constitute management. However, when most people talk about "management," they imply the activities of executives who are responsible for making decisions which affect large groups of people, or which influence the spending of large sums of money. Many engineers function as managers when they are active as project engineers, chief engineers, branch or district engineers in agencies, and at other levels of management characterized by supervising small numbers of their fellow workers. Many of these only consider themselves "managers" when they feel forced to abandon their own engineering and technical work to devote most of their time to making major decisions concerning the allocation of labor and materials in accomplishing overall project objectives.

Since managers are considered to be those who make important decisions concerning the employment of large numbers of people or the expenditure of large sums of money, it is obvious that management is a very important and exacting activity. Why would engineers be attracted to management roles? One of the most obvious reasons is that, in general, managers and executives are richly rewarded in terms of salary and other benefits. Typically, the chief executive of a corporation will be rewarded with a salary three to four times as high as the highest paid engineer employed by that corporation. Many times, upper-level managers in corporations or in companies may be rewarded with a share in the profits or with part ownership of the company or corporation.

Another major attraction to a management career is the opportunity to make important decisions. Many individuals seek challenge in life and feel rewarded when they are given the opportunity to make decisions affecting large numbers of persons or large sums of money. It has been stated in many articles and books that the most obvious attractions to management careers are the power associated with management positions and the individual status that a person achieves with such power. However, it is an over-simplification to say that most engineers who enter management do so simply to achieve status. It is much more meaningful to observe that most managers are motivated by the challenge of making important decisions. In making these decisions, managers must make full use of their technical resources and also their ability to utilize fellow workers to their greatest potential. They must be creative and imaginative in developing new solutions to old problems. In all of these activities, managers who possess the proper perspective will feel a great sense of achievement and accomplishment.

On the other hand, certain individuals will find careers in management unappealing because of some of the disadvantages inherent in management activities. One of the most serious disadvantages of a management career is the high level of emotional stress under which many executives often work. This emotional stress may arise from the necessity to make important decisions, or it may be created by the need for a manager to hold her or his own emotions in check in order to maintain positive working relationships with colleagues and subordinates.

Before a person embarks on a career in management, it would be wise to consider some of the demands made on managers. The successful manager must be capable of leading and motivating people, both technical and non-technical. The manager must desire to be in a decision-making position, and must feel comfortable in the control of the destinies of fellow workers and of customers or clients affected by her or his activities. In general, the successful manager must show a willingness to place the interests of the organization ahead of her or his own interests. He or she must be willing to spend long hours working toward her or his goals, and should be motivated by a strong desire for personal success. Since the rise to top level management may not be rapid, he or she must be capable of enduring some disappointments along the way and should be optimistic about the chances eventually to succeed. Obviously, in order to succeed, he or she must be able to persevere in accomplishing a given task and must be capable of exercising good judgment in making basic decisions. Additionally, he or she must be able to convey to colleagues her or his ideas on management, her or his decisions and choices between alternatives, and the degree to which her or his decisions may be superior to those made by others.

If an individual who is interested in the field of civil engineering examines all of the demands associated with a management career, he or she will find that an education in civil engineering will serve as an excellent foundation for an initial career in the technical field of civil engineering followed by a subsequent career in management. Certainly, this is borne out by the large number of civil engineers who are successful managers.

CONSULTING ENGINEERING

This is a good point at which to discuss the opportunities for civil engineers to be self-employed as consultants. Consulting engineering is a form of civil engineering practice which includes all of the functional specialties mentioned above. Basically, a civil engineer who wishes to be a consultant must be a registered professional engineer in the state in which he or she offers services to clients.

Many civil engineers pursue consulting activities of a general nature. Their day-to-day activities may include the gathering of basic data such as that obtained in a land survey, the measurement of traffic at a particular location, or the determination of the flow in a particular waterway. Consulting civil engineers are often called upon to develop plans and designs on the basis of the data they have gathered or on that which has been gathered by other engineers.

The consulting civil engineer may be also engaged in any one of the particular technical specializations which were mentioned earlier in this chapter. Many civil consulting engineers are engaged in surveying activities, in structural planning and design, in design of transportation systems, in geotechnical engineering, in environmental engineering, in urban planning, and in public works consulting capacities. Consulting engineers in some areas of technical specialization at times may be engaged in research and development activities, in gathering basic data for planning purposes, in formulating designs for specific projects, in representing an owner or a client as a supervisor on a particular project, or in exercising a management role in coordinating the activities of a number of other engineers.

Obviously, if a consultant is successful and wishes to expand her or his activities, he or she may employ other engineers who then assume portions of the work load. With continued success, the consulting engineer will assume the role of a manager in supervising the work of her or his associates. If the consulting firm grows to the point where it has large financial resources and employs considerable numbers of personnel, the consulting engineer may be found to be functioning primarily as an executive with little time devoted to actual engineering work of her or his own.

CONCLUSION

The vast spectrum of activities which are available to a person considering a career in civil engineering should be obvious from the descriptions contained in this chapter. When pursuing a career in any one of the technical areas and at any one of the levels of functional specialization described earlier, the civil engineer can look forward to adequate compensation in the form of a high salary, excellent fringe benefits, and relatively stable employment conditions, which are described in the next chapter. Most importantly, he or she can look forward to a high degree of satisfaction.

CHAPTER 4

JOB DUTIES, SALARIES, WORKING CONDITIONS

In Chapter 3 we described the many specializations within the field of civil engineering. In Chapter 2 the various types of educational programs available in civil engineering were described. However, we may not have made it absolutely clear exactly what type of work you could expect to do in any of the special fields or any of the special areas in civil engineering. In this chapter we hope to clarify the possible job duties for a few of the areas of civil engineering. This description is intended to indicate the nature of the work that the civil engineer would encounter at various stages during a professional career.

Finding the right type of work, work that is both a challenge and an opportunity for advancement, is more important for a recent graduate than getting a high starting salary. Salary almost never appears at the top of the list of answers when engineers are asked to indicate those items that they consider important in the ideal job. Usually, salaries show up in third or fourth place. The great majority of practicing engineers consider that the most desirable characteristics about a job are: 1) opportunity for advancement; 2) creative, challenging work; 3) good salary; 4) recognition of achievement; and 5) the chance to keep up with new developments in the field.

This ranking should not be interpreted to mean that engineers have little interest in money. What this ranking really means is that for most engineers, it has not been a serious problem to consider what kind of salary they will receive. In other words, most civil engineers simply have not worried about salaries because they have been in a "seller's market" since about 1950. Of course, not all branches of engineering receive the same average salaries, as shown in Chapter 1. However, on the average, salaries paid to engineers are significantly higher than salaries received by most other professionals. Again, on the average, civil engineers' salaries tend to be lower initially than the salaries paid to members of other branches of engineering. However, the difference in average salaries is a small percentage of the salary itself. Additionally, a number of civil engineers are among the most highly paid engineers in the United States. These engineers are those who own or who are partners in private consulting or construction companies. We will give detailed information about typical salaries in this chapter, but first we will talk about some of the more important job characteristics such as the nature of the work you would be doing as a civil engineer.

SPECIFIC JOB DUTIES

In civil engineering today, in all of the various fields described in Chapter 3, the types of job activities available are quite varied. The many areas of specialization in civil engineering, and the different levels of responsibility on an engineering project, create a situation in which an individual's particular responsibilities could range over a wide variety of tasks. To some degree, the level of experience of the individual will determine the responsibilities assigned. The young engineer with little experience usually will start out with minor to moderate responsibility. As the individual gains experience, he or she will be assigned more important decisions to make.

At the lower levels of engineering responsibility, for example, the engineer may be assigned to work with surveyors in the precise layout of facilities during construction. On the other hand, a junior engineer may be engaged in the development of a small portion of the plans or designs for the construction of a major facility. A beginning designer may be assigned the task to select the most economical materials to use in a particular phase of a construction project. During the course of a design effort, the junior engineer may be assigned the task to calculate the quantities of materials and labor involved in a particular design and to estimate the cost to carry out that design. In general, for the engineer with little experience, the degree of responsibility assigned will be minor. Usually the junior engineer is assigned to accomplish a small part of a large task or all of a relatively minor design or plan.

The particular tasks to which the engineer is assigned will vary also with the area in which the engineer is specializing. For example, a structural engineer will work to estimate or determine the loads to which a structure will be exposed. In estimating these loads, the engineer will calculate the weight of all the building materials as well as the weight of snow which could accumulate on the roof of the finished building. He or she will also make an attempt to calculate the weight of all the people and equipment which could be placed inside the building at any time. He or she also must consider the effect of wind loads on the sides and top of the building as well as other forces such as earthquake forces. After determining all of the various loads which could be placed on the building, he or she must calculate the corresponding forces which are created in all of the members of the building structure. In making this calculation, a considerable amount of judgment must be exercised concerning the probabilities of the loads actually being present at the same time on the structure. In other words, he or she will have to decide if it is probable that a maximum snow load will exist on a roof of a building at the same time that high-velocity winds strike the sides of the building. He or she will have to decide if earthquake forces should be combined with wind load and snow load in the design. This decision will require a considerable amount of experience, and usually only senior level engineers will make this sort of judgment. After the structural engineer has developed a combination of forces which he or she considers most appropriate for a particular building, he or she will have to select the most efficient combination of materials and structural shapes to carry the loading.

As another example, the water resources engineer in a typical project would be required to estimate the demand by a community for water supplies and for facilities to carry away and treat wastewaters.

A junior engineer could be employed in estimating these quantities. A more senior engineer usually would be in charge of designing the actual facilities for treatment of water and water distribution. A senior engineer also would be in charge of the design and planning of a sewer system or a wastewater treatment plant. Usually, engineers with little experience would be assigned only small portions of the overall design responsibility.

Some of the studies undertaken by civil engineers can become extremely complex and sophisticated when they are called upon to forecast population trends and developments. Urban planners and civil engineers working in municipal engineering often must estimate future populations and trends in the growth and development of suburban areas around cities. In these growing areas, the civil engineer will be required to plan and provide all of the required public services in advance of the time when the people move into the new areas of the community and need the services. In this way, the civil engineer involved in public works must anticipate and be responsive to complex social needs in the community. A major area of responsibility for the civil engineer is the study of the relationship of her or his activities to the quality of the environment. Civil engineers are responsible for the planning and design of facilities intended to control water and air pollution. Civil engineers also are called upon to prepare or contribute to environmental impact statements which are developed to quantify the environmental effects of proposed engineering works.

In summary, the variety of job duties today is so wide for civil engineers that no new civil engineering graduate should worry about finding a job that would match her or his interest, educational background, and level of experience. Civil engineering activities range from the rehabilitation of older cities, to the development of new transportation networks, to the construction of orbiting space stations. The young engineer should expect to get only "a piece of the action." However, with more experience the engineer's responsibilities will be extended.

Since the new civil engineering graduate has only a limited amount of professional experience, the responsibility given on the first few assignments will be somewhat limited. However, most employers attempt to develop the young engineer professionally as rapidly as possible. Rapid professional development of the engineer makes that individual much more valuable to any employer. The engineer becomes much more valuable when able to accept responsibility for major engineering decisions.

Usually, of necessity, the beginning engineer will start her or his career making decisions which can be based on academic experience rather than professional experience. The beginning engineer may find herself or himself in charge of a surveying party. As we have said, a recent graduate may be assigned the responsibility for material testing as a field inspector on a construction site. The recent graduate may be assigned to estimate the costs for materials and labor on small or medium construction contracts. If the new engineer accepts a position with a private consulting firm, more responsibilities may be given sooner than if he or she elects employment with a governmental agency. Ultimately, however, the engineer working for a governmental agency may be called upon to make much more important decisions. In most instances, the new engineer will work closely with, and under the supervision of, a more experienced engineer. The engineer working with a consulting firm usually will be employed on some facet of the design or planning of a facility such as a bridge, a building, a highway, or a water supply or pollution control facility.

A recent graduate who goes to work with a city or county government will deal almost at once with the public. In these types of positions, the civil engineer must deal with technical aspects of problems but also must confront social and political situations. For example, he or she may be called upon to give technical assistance to a zoning commission in a city or county and will be required to predict the social and environmental effects of proposed activities, as well as to evaluate the technical efforts involved. The young engineer may be assigned the task of being a liaison between the governmental agency and a contractor who holds a contract to perform an engineering task for the city or county. Obviously, these engineers will be working with elected officials who are charged by the public with the responsibility of providing services and facilities. The young engineer may be assigned to meet with neighborhood groups to explain the governmental agency's position with respect to some decision on an engineering improvement or project.

More often, the engineer with little experience will be assigned the tasks of gathering and assembling the engineering data needed to allow a more experienced professional to make a major technical decision or complete a complex design. Carrying out all of these minor assignments will help the young engineer to gain the experience necessary to make major decisions at a later time.

When the engineer has gained experience, he or she can expect added responsibilities and should be dissatisfied if not assigned more important tasks. Of course, he or she can also expect to receive added compensation and recognition from the employer.

For the professional civil engineer, the ability and willingness to accept increasing amounts of responsibility is one of the most important aspects of her or his professional career. The civil engineer is constantly dealing with matters which directly affect the health and safety of the general public. Obviously, incompetence in such a position of responsibility cannot be tolerated. On the other hand, when an individual accepts such responsibilities, he or she should be amply and appropriately rewarded and compensated. Wages and salaries, as well as other forms of compensation, should be based upon the individual's performance of job assignments and fulfillment of responsibilities.

WAGES AND SALARIES

The young civil engineering graduate can be sure of receiving a salary consistent with the education required to obtain the degree. Salary also should be proportional to the importance of the work that is assigned. The rate at which an engineer just out of college advances in salary for the first several years of employment is to a great extent independent of performance, unless that performance is unacceptable and he or she is released. Usually, if the individual is considered sufficiently valuable to stay on the payroll, salary increases in the

Table 1

STARTING SALARIES
(monthly rates)

Year	Engr	Sales	Acctg	Lib Arts	Chem	Econ	Business
1956	$415	$370	$372	—	—	—	$363
1960	510	440	446	—	—	—	427
1965	632	521	561	—	—	—	520
1970	872	715	840	682	809	740	677
1975	1062	862	990	776	992	851	812
1985	2265	1548	1697	1503	1897	1792	1636

Sources: 1956-1975 data; Frank S. Endicott, Director of Placement, Northwestern University. 1985 data; College Placement Council, CPC Salary Survey, July, 1985.

first year or two will be practically automatic. After this probationary period, the quality of the individual's performance will begin to affect salary more directly. A truly outstanding person will move ahead very rapidly. Of course, the actual salary which the person receives will depend to a great extent upon what sector of business or government he or she has selected for employment. Civil engineers employed in various sectors of society do not all receive the same salary. Nevertheless, a competent civil engineer can expect more or less steady salary growth for a period of as long as five to ten years after beginning employment. Remember that the civil engineer's starting salary is much higher than that of many other professionals and many other occupational groups. The steady salary growth with experience coupled with the high starting salary makes engineers' compensation generally much higher than other occupational fields.

The starting salaries paid to recent engineering graduates have consistently been higher than starting salaries paid in many other occupational fields. Table 1 shows starting salaries for a number of occupations, including engineering, for a 30-year period between 1956 and 1985. Starting salaries for the various branches of engineering are shown in Table 2. The average salary paid to beginning civil engineering graduates is approximately 86 percent of the overall average of beginning starting salaries of all engineers. Figure 1 shows CE starting salaries. It is obvious that, *on the average,* beginning civil engineers are paid slightly lower salaries than the average for all beginning engineering employees. However, starting salaries do not reflect the entire picture. Not only do salaries increase very significantly with the experience an engineer gains, but in some fields, the engineer's salary increases very dramatically if he or she owns her or his own company or is a partner in a consulting organization. This is particularly true in civil engineering. Before we go on to talk about the influence of experience on salary, however, another important point can be made concerning starting salaries. Table 3 shows starting salaries paid in a number of fields to men and women. As mentioned in Chapter 1, the future holds great promise for minorities and women in engineering. Table 3 shows conclusively that women command equal or higher salaries in almost all of the engineering fields. Only in mining engineering and in nuclear engineering were starting salaries lower in 1985 for women than for men. This situation is in contrast to the general situation in sales, economics, chemistry, computer science, the health professions, and mathematics, as shown in Table 3. In all of those other fields, starting salaries for women in 1985 were lower than starting salaries for men. Among all of the branches of engineering, only in mining engineering were the starting salaries for women significantly lower than the starting salaries for men.

Table 2

ENGINEERING STARTING SALARIES
(monthly rates)

July, 1985, Data

Branch of Engineering	Monthly Salary
Chemical	$2,369
Civil	1,969
Electrical*	2,283
Industrial	2,191
Mechanical	2,259
Mining	2,190
Nuclear	2,283
Overall Average	$2,265

*Includes Computer Engineering

Source: College Placement Council, CPC Salary Survey

As mentioned previously, engineering salaries also greatly increase with increasing experience. Figure 1 shows the increase in annual salary for all engineers employed in all industries in the United States, with the number of years of experience since the engineer received the bachelor's degree. The solid dark line in the figure represents the average salary figures. The long dashed lines represent the upper one-fourth of all of the salary figures and the lower one-fourth of all of the salary figures, respectively. The short dashed lines represent the upper ten percent and the lower ten percent of all of the salary numbers. This figure shows obviously that for a very small minority of engineers, increasing experience brings only a modest increase in salary. Beginning salaries shown in this figure varied from about $22,000 up to about $30,000 per year. In other words, only about ten percent of the engineers starting work in 1985 received starting salaries lower than about $22,000 a year. On the other hand, only about ten percent of all engineers beginning work in 1985 received starting salaries higher than about $30,000 a year. The average starting salary in 1985 was approximately $26,000. In contrast, the average salary for engineers with 33 or more years of experience was $46,000 a year in 1985. Ten percent of the engineers with 33 or more years experience had salaries greater than $67,000 a year in 1985, while ten percent of the engineers with 33 or more years experience had salaries below $34,000 per year in 1985. Figure 1 clearly shows the important influence of experience on engineering salary and also shows very dramatically the important influence of years of experience for the very successful engineer.

Table 3

JULY 1985 STARTING SALARIES

Field	Men	Women
Accounting	$1,697	$1,698
Sales	1,604	1,486
Economics	1,752	1,689
Engineering		
Chemical	2,367	2,372
Civil	1,956	2,024
Electrical	2,282	2,294
Industrial	2,174	2,217
Mechanical	2,253	2,296
Mining	2,203	1,871
Nuclear	2,286	2,243
Chemistry	1,952	1,848
Computer Science	2,102	2,055
Health Professions	2,063	1,722
Mathematics	2,062	2,032

Source: College Placement Council, CPC Salary Survey

Figure 1

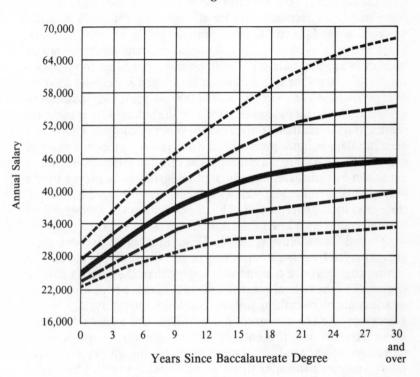

Source: *Professional Income of Engineers,* Engineering Manpower Commission, 1985.

━━━━ Average salary figures

━ ━ ━ Upper and lower 25% of salary figures

----- Upper and lower 10% of salary figures

The starting salary paid to civil engineers (and all other engineers) varies significantly with the employment sector selected by the beginning engineer. The average starting salary for engineers employed with the federal government in 1985 was slightly above $22,000, as compared to an average salary for all beginning engineers in all industries in the United States of approximately $26,000 per year. Engineers employed with the federal government for 25 years had salaries of approximately $37,500, on the average. In comparison, the average salary for all engineers in the United States with 25 years of experience in 1985 was almost $44,000 per year. This indicates in general that salaries paid to engineers by the federal government are significantly lower than salaries paid by other employers. This difference in salary is more pronounced with greater number of years of experience. Also there is a much greater range in salary for engineers employed by industries as compared to engineers employed by the federal government. Many civil engineers are employed by federal, state, and local governments. This is one reason the average salary for civil engineers is lower than the average salary for all engineers in the United States. However, a significant number of civil engineers are employed by private consulting firms. The salaries paid to consulting engineers are considerably higher than the salaries paid to engineers by the federal government, and there is a wide range in the salaries paid by consulting firms to engineers. On the average, an engineer employed by a consulting firm will receive higher salary after about ten years with that consulting firm than if he or she were employed by another industrial employer. The top ten percent of the salaries paid by consulting firms to engineers is virtually identical to the top ten percent of salaries paid to all engineers by all industries in the United States. Construction engineers are among the most highly paid engineers in the United States. An engineer with 25 years of experience in the construction industry, on the average, in 1985 was paid a salary of almost $64,000 per year. This is especially significant to civil engineers since most of the engineers employed in construction activities are civil engineers.

It is true that, on the average, civil engineering starting salaries are lower than the starting salaries paid to many other engineers. One reason this is true is that many civil engineers begin their careers working for governmental agencies. A large number of civil engineers remain with such agencies throughout their entire careers, in many cases because of the challenging nature of the work and the close contact with the public which is fundamental to the type of engineering activities associated with public works. A smaller number of civil engineers work for consulting firms at significantly higher salaries, on the average, than the salaries paid to government engineers or consulting engineers. In other words, a number of civil engineers work for salaries which are slightly below the average salaries paid to all engineers in the United States. However, another number of civil engineers work for salaries which are significantly higher than those paid to most all other engineers. This situation is exaggerated as the engineer gains more and more experience.

FRINGE BENEFITS

The direct salary paid to a civil engineer is one of the factors considered very seriously by the engineer in evaluating an employment offer. However, the salary alone is not the total compensation. Total compensation includes direct salary plus any fringe benefits offered. The fringe benefits may be just as important as the salary. Every job offer, without exception, offers some type of fringe benefits package. All of the fringe benefits are important to evaluate in examining a job offer. Among the more common benefits that you should expect to receive as a civil engineer are paid vacations, holidays, sick leave, retirement pay, and workmen's compensation insurance. In addition to these benefits, many firms and governmental agencies will have other fringe benefits to offer. These can include military leave or jury duty pay. Other fringe benefits include incentive compensation, extra salary or bonuses paid to the employee as a result of a successful year experienced by a company, or the successful completion of a project. This is particularly true in the consulting engineering field and in the construction industry. Early completion of a contract often produces a bonus payment to the successful contractor. This bonus payment will produce bonus salaries for the engineers working for the contractor. Other fringe benefits may include payment by the employer of the membership fees for professional societies or the fees for professional licenses. Many engineering employers send their civil engineering employees to meetings, conventions, and short courses at company expense. Many companies and agencies pay for life insurance and medical insurance for many of their employees. Some companies will pay educational costs for their professional employees. Many people are familiar with fringe benefits such as the free use of company vehicles or company-owned recreational facilities. A much more important fringe benefit is profit sharing. Profit sharing means that when a company is particularly successful, the profits that are made are shared out in one of several ways with the employees. In many cases, the profit shares are more significant for professional employees, and especially for upper level engineers. Profit sharing can include employee ownership in a company. In other words, in addition to giving an employee a bonus payment after a successful year or after a completion of a successful contract, the employee may be rewarded with shares of company stock. This type of fringe benefit has significant long-term considerations. Owning part of the company which employs her or him may be a powerful incentive to the engineer to work harder and be more careful in assignments. Also, the successful completion of an assigned task may bring added satisfaction to the engineer if the engineer is a partial owner of the company. Depending upon the particular package of fringe benefits offered to the individual, the total compensation may vary from as low as 125 percent to more than 200 percent of the direct salary paid to the engineer. The engineer must give careful consideration to the fringe benefits offered by a firm or agency before accepting a job offer. Of course, in many cases, a very attractive fringe benefits package may be offered with a lower direct salary, and the engineer must be very careful to consider the long-term benefits associated with both the direct salary and the fringe benefits package.

GENERAL EMPLOYMENT CONDITIONS

Since the civil engineer is a professional employee, he or she will find that the general conditions of employment will be quite different from those of a tradesperson, a vocational worker, or other non-professional employee. Usually the professional employee does not use a time clock to record arrival and departure from the office. A time clock is particularly inappropriate for most engineers, since situations will arise which may require unusual effort and late hours, as well as many trips in and out of the office during the course of a working day. In general, unless extra work and overtime occur on a regular basis, the civil engineer will not be compensated directly for this extra work. However, indirect compensation may come in the form of being able to report late to work the day after a late night assignment or the opportunity to leave early on a weekend in return for an extra effort late at night earlier in the week.

The professional employee usually is continually informed about the employer's objectives, policies, and goals. In many cases the engineer will be a part of the firm's management and may even be a partial owner of the company. Almost all professional employees will be provided with a well-appointed office, adequate support staff, and physical facilities which will promote personal efficiency. For engineers, duties and levels of responsibility will be clearly defined by the employer and usually will be reflected in the engineer's position and title. Because of this professional position of employment, the civil engineer will have obligations to the employer over and above those of the non-professional employee. The engineer must accept assignments only when absolutely sure he or she is qualified to carry out the work. The engineer must be diligent, competent, and honest in completing assigned duties. The professional responsibilities of the engineer require also trying to contribute creative and resourceful ideas to solve the problems encountered in the course of employment. The foremost obligation, however, for the professional employee is due regard for the safety, life, and health of the public and fellow employees.

The professional employee must be careful to avoid any conflict of interest. For example, it would be unethical for a civil engineer to work at night for a competitor of the employer for whom he or she works during the day. It would also be unethical for an engineer to take information learned during the course of regular assignments and disclose that information to a competitor or to use that information for personal gain. The engineer cannot disclose any technical, proprietary information that would have any adverse effects on her or his employer or firm. If the engineer discovers that some action of her or his employer is contrary to the public interest, the engineer is obligated to inform the employer first about the problem and give the employer the opportunity to correct the situation. If the employer does not correct the situation, and public welfare continues to be in danger, the engineer must disclose the situation to the public. On the other hand, the engineer has the right to expect professional responsibility from colleagues and from the employer. If the engineer discovers a circumstance dangerous to public health and welfare, he or she has the right to expect that the employer would make every effort to correct the dangerous situation. Just as the engineer should show loyalty to colleagues and the employer, he or she has the right to expect similar loyalty in return. The common bond which unites engineers in their efforts to improve the quality of life is one of the most important intangible factors which leads to professional and personal satisfaction for the civil engineer.

OVERALL DEMAND FOR CIVIL ENGINEERS

Since the beginning of World War II, technological change in the United States has taken place at an ever-increasing rate. This rapid pace of technical change has produced changes in the style and standard of living throughout the United States and Western Europe. Increasing emphasis and dependence on technology has produced an increase in the demand for engineers which has been relatively constant since the early 1940s. This strong demand for all types of engineers, on the average, has not held true for all branches of engineering during all of the past 40 years. The fluctuations in the demand for civil engineers, for example, have already been mentioned in Chapter 1. A high percentage of all the students enrolled in engineering curricula in 1975, for example, were civil engineering students. This high enrollment reflected a very strong demand for civil engineers in the mid-1970s for work in pollution control, energy supply, and environmental protection. In the early 1980s, in contrast, a national emphasis on defense and computer technology led to a much lower percentage of students enrolling in civil engineering programs than in other fields. Ironically, the demand for civil engineering graduates has remained relatively steady. What this means is that there have always been jobs for civil engineering graduates, but during the 1970s when civil engineering enrollments accounted for a higher percentage of the total engineering enrollment, the graduating civil engineers had fewer job offers to choose among than will the civil engineering students who graduate in the late 1980s. Since a lower percentage of students have enrolled in civil engineering in the early 1980s, there will be fewer graduates to fill the jobs opening in the late 1980s and early 1990s. This situation will be very advantageous for the students who graduate at those times. The most important factor to consider, however, is that there has always been a steady demand for civil engineering graduates. There is very little likelihood that this demand will ever disappear. The types of activities in which civil engineers are engaged are critical to the continuing life of the United States. These activities range from the most fundamental, such as the supply of drinking water, to the most sophisticated, such as the construction of space stations. The decrease in spending for public works (transportation systems, sewer systems, water supply systems, etc.) during the early 1980s has simply created a serious problem which will face coming generations who will be required to repair and renovate these facilities. Such repair and renovation obviously will require civil engineers. "The bottom line is," as people on Madison Avenue say, the employment prospects for civil engineers in the past have been good to excellent, they are good at the present, and in the future they should become even better.

A civil engineer inspects plans at a construction site. (American Society of Civil Engineers photo)

CHAPTER 5

EMPLOYMENT AND ADVANCEMENT

The general working conditions, salaries, and benefits for civil engineers were described in the last chapter. In this chapter, the stability of employment in the field of civil engineering will be described. Also, the steps to be taken by the recent civil engineering graduate in finding employment will be described.

In general, the unemployment rate for civil engineers is extremely low, and the new graduate will find a steady demand for her or his services. It will be much more important for that graduate to choose the job most appropriate to her or his desires and capabilities rather than to worry about unemployment. After the young civil engineer has selected a job and has been employed for a time, there are various paths that he or she can follow to achieve professional advancement. Those paths to career advancement also will be described in this chapter.

JOBS

In any discussion of employment and career advancement, one of the most important considerations is job stability. In other words, is there likely to be a variable demand for individuals engaged in a particular type of activity so that periods of high salary and many job offers alternate with periods of unemployment? A great deal has been written and spoken about the stability of the job market in engineering since the late 1960s when aerospace engineers were hard pressed to find work. There has been continuing controversy over whether or not there will be enough engineers to meet projected demands. Because the employment picture does vary with economic conditions, engineering enrollments tend to show some fluctuation. Shortly after World War II, with the return of many servicemen to schools and colleges in the United States, engineering enrollments reached a peak, with more than 80,000 freshmen enrolling in engineering courses in 1946. In the five years from 1946 to 1950, however, freshmen enrollments dropped to approximately 35,000. What caused this rapid and serious decline in the number of freshmen college students choosing engineering as a career?

Shortly after World War II, the federal government released statistical analyses of the engineering profession, including predictions for jobs in the coming years. These predictions included a warning that, with the anticipated enrollments of more students in engineering courses (at the 80,000 per year level), a surplus of engineers soon would be created in the United States. This prediction was given great publicity in newspapers and magazines. As a result, many high school seniors decided to enter other careers, and freshmen enrollments in engineering dropped accordingly. Then, in the early 1950s, with the start of the Korean War and with a corresponding increased need for engineers and technically trained specialists, freshmen enrollments in engineering once again increased. By 1956, freshmen enrollments in engineering again reached the level of about 80,000, equaling the 1946 peak.

During these years, continued emphasis was given to the need for scientists and engineers by the growing competition between the United States and the Soviet Union, particularly in aerospace activities. Again, however, the rapid increase in freshmen enrollments led to predictions that the job market would be flooded with graduate engineers in the early 1960s. As a result of these predictions, freshman engineering enrollments once again declined to a level of about 65,000 in 1962 and 1963.

Of course, these changes in the numbers of college freshmen enrolling in engineering courses were reflected in similar increases and decreases in the number of persons graduating from engineering curricula. For example, the peak enrollment of over 80,000 freshmen in engineering in 1946 yielded more than 50,000 engineering graduates in 1950. The rapid decrease in freshmen enrollments in engineering, with a low of about 35,000 in 1950, produced a corresponding decrease in the number of engineering degrees awarded four years later, with a low of approximately 25,000 bachelor's degrees in engineering being granted in 1954 and 1955.

The mid-1960s saw a significant increase in the number of freshmen enrolling in engineering curricula, primarily as a result of the greatly expanded requirements for engineers in aerospace activities. The federal government produced data which showed that the demand for engineers and scientists in 1965 and 1966 was more than twice as great as the demand in 1960. This demand produced rises in enrollments which, in turn, resulted in a steady increase in the number of engineering graduates from 1965 to 1970.

The economic recession and the decrease in demand for engineers in the late 1960s and the early 1970s was again publicized widely. A large proportion of the engineers and scientists who were suddenly unemployed had been active in the aerospace and defense programs. Some of the most serious cutbacks in the aerospace and defense industries were experienced at the executive level of management and at the senior scientist/engineer level. As a result, many of the newly unemployed engineers were persons with years of experience and advanced degrees, particularly Ph.D.s. The situation of Ph.D. engineers with 15 years experience being suddenly out of work, or employed as janitors and taxi drivers, made excellent material for television news programs, newspapers, and magazine articles. As one would expect, freshmen enrollments in engineering correspondingly dropped to approximately 50,000 in 1973, producing a significant reduction in the number of engineering graduates in the middle 1970s. Since that time, freshmen enrollments have increased and also stabilized such that neither a great shortage nor a glut of graduates is anticipated in the foreseeable future.

Can a person make any sense out of these recurrent swings in demand and supply? It is possible to clear up a considerable amount of the confusion by focusing attention not on the number of students in engineering courses and the number of graduates from engineering schools but on the number of engineers who could *not* find suitable employment during these same times.

The United States Bureau of Labor Statistics compiles data on the unemployment rates for a variety of professions. The Bureau data on unemployment rates for engineers show that the maximum unemployment rate in the ten years from 1963 to 1973, for engineers, was slightly more than three percent in early 1971. This unemployment rate was approximately half the rate for all workers in the United States at this time. During "the best of times" (1963-69), the average unemployment rate was only one percent, with the highest rate being slightly more than two percent in early 1964 and the lowest being less than 0.5 percent in late 1966. Since the great decline in the demand for aerospace and defense engineers in the early 1970s, the unemployment rate for engineers dropped from the all-time high of 3.2 percent in 1971 to a level of approximately 1 to 1½ percent in the mid 1970s. Thus, it appears that even when considerable nationwide attention was given to the soft employment picture for engineers, their unemployment rate was *less than* half the unemployment rate for all workers in the United States.

Specifically, what can be reported about the unemployment rate for civil engineers? In general, the unemployment rate, associated with fluctuations in supply and demand, varies much less from year to year for civil engineers than for engineers in other branches. This is true for several reasons. First, civil engineers are engaged in activities which for the most part are associated with the necessary day-to-day activities of today's society. The various engineering specializations which were described in Chapter 3 correspond to activities necessary for the health and welfare of the population. For example, it is hard to imagine that a civil engineer engaged in obtaining and distributing water supplies to a large American city suddenly would be unemployed as a result of a minor economic recession. Civil engineers engaged in other forms of public works activities such as solid waste management, wastewater treatment, traffic control, and other similar activities are hardly candidates for being among the unemployed even during periods of economic depression.

A second factor which helps to prevent any drastic fluctuations in the demand for civil engineers is the fact that in times of economic depression, the federal government in the United States traditionally has attempted to use some public works programs to help revitalize the national economy. Of course, these activities must be planned, designed, and supervised by civil engineers. As a consequence, civil engineers are in demand even when some of their professional colleagues in architecture or other branches of engineering may have difficulty finding suitable jobs.

A third factor tending to lead to a more constant demand for civil engineers, when compared to that for other types of engineers, is the fact that many civil engineers are employed by federal, state, and local governments and are not subject to the hiring and firing policies of industrial corporations. In general, employment with governmental agencies is much steadier and more secure than employment with private corporations.

Finally, another factor in the relatively steady demand for civil engineers arises from the fact that they are generally able to find employment in all parts of the world. Civil engineers, in contrast to other types of engineers whose activities are directly related to industrial societies, may be employed in "developing" nations on public works projects designed to improve living conditions in those countries. In times when market fluctuations restrict industrial output, and possibly lead to the unemployment of engineers associated with manufacturing and industrial production, civil engineers will find relatively steady employment since their activities are related to long-term goals of improving living conditions; developing water supply, power, and transportation systems; and constructing the structures and facilities related to such systems.

Since this brief analysis of the job market indicates that civil engineering graduates will find little difficulty in obtaining employment, it appears that it may be more important for a person considering a career in civil engineering to think about choosing the right job rather than being concerned with whether or not a job would be available. The next section describes that process.

CHOOSING A JOB

The prospective graduate of a civil engineering curriculum is likely to be confronted with a considerable list of job offers and opportunities in the weeks before graduation. Most of these job offers will appear to be worthwhile and relatively attractive. This is logical, since recruiters who visit college campuses attempt to present their companies and agencies in as favorable a light as possible. As a consequence, the new civil engineer may be faced with a serious problem in selecting that one job which will best suit her or his individual capabilities and desires.

In order to select that one "best" job, the prospective graduate and even the experienced civil engineer seeking a new job should evaluate several major factors concerning job offers. These factors include: personal preferences; the professional reputation of the prospective employer; the opportunities for career advancement, including continued education in the jobs offered; and the salary, fringe benefits, and job security conditions which accompany a particular job offer.

PERSONAL PREFERENCE

The most important factor for the civil engineer to consider in evaluating a job offer is her or his personal reaction to the description of the work and duties which would be assigned. In other words, if he or she hopes to be successful in that job, it must have a personal appeal.

One would expect that the job offer will be in the field of civil engineering; however, occasionally employers will approach civil engineering graduates with offers for work in other fields of activity such as sales or junior level management. Some individuals may find such offers attractive and may be successful in applying their personal capabilities and engineering training in such a field. However, this is the exception rather than the norm. The typical new civil engineering graduate would be well-advised to select a job in the field of civil engineering in an initial phase of her or his career rather than be persuaded to enter another field on the basis of higher pay or similar inducements. After gaining some experience in a particular job assignment, the new engineer will be better prepared to undertake challenges in other fields such as sales. If the new graduate chooses to enter a completely different field of endeavor such as sales or management, there is a greater chance of being dissatisfied with the first work assignment than if he or she were to accept a technical civil engineering job, since most of her or his college training has prepared her or him to meet that type of challenge. If the graduate undertakes a job assignment in which he or she is unhappy, he or she is likely to be an unsuccessful employee, regardless of the field. Consequently, the chances for advancement in that field will be minimal.

The new civil engineering graduate should also evaluate the conditions of work associated with any particular job offer. If he or she is fond of working and living in the outdoors, he or she might be quite well pleased with the work of a construction engineer. However, if he or she prefers to work in an office or in a research environment, construction supervision at a remote location would likely be unattractive. The geographical location of the job assignment, whether it is in a large metropolitan area or in a smaller community, can be an asset or a liability to a particular job, depending upon the desires of the individual taking that job.

Another important factor to consider in evaluating a job offer is the element of challenge associated with the prospective job. Challenge is an important ingredient in any job and the key to professional fulfillment. Perhaps the greatest single source of dissatisfaction among recently hired engineers has proven to be boredom and a sense of unimportance of the assigned duties associated with the initial engineering assignment. In some instances, this boredom is the result of the reluctance of employers to entrust new engineers with important decisions; as a result, recent graduates sometimes are given routine or simple tasks which involve no important decision-making. It is important, in evaluating job offers, to attempt to determine the responsibilities and duties which initially will be assigned, and to determine if those responsibilities constitute a sufficient challenge to the individual's capabilities.

OPPORTUNITIES FOR WOMEN AND MINORITIES

In the past, members of minority groups have occasionally been led to believe that they would be the victims of tokenism in seeking employment as engineers. In other words, they were given jobs in various agencies or corporations merely so that those agencies or corporations would have a minimum number of minority group members. This situation may exist in some areas but certainly does not exist at the present time in the field of civil engineering.

Several reasons can be cited to account for the increasing availability of challenging job opportunities for women and minorities. Obviously, the civil rights movement which created sweeping social changes in the United States in the 1960s had great effects in the field of engineering, as it did in other professions. Not only were professional organizations such as the National Society of Professional Engineers and the American Society of Civil Engineers made aware of the lack of equal opportunities for minorities, but engineering educators also were urged to exert greater efforts in the area of minority recruitment.

In addition, women are finding increasingly important opportunities in civil engineering careers as a result of the changes in the attitudes on the part of both engineers and employers. At the same time, the work of the civil engineer has changed significantly in the last 50 years. Many activities which formerly were carried out by civil engineers in the field, in the mine, or in the quarry have been relocated to more pleasant surroundings. The percentages of civil engineers engaged in planning and design activities have increased in comparison to the number of civil engineers engaged in construction supervision and similar field work. A woman is no longer at a disadvantage in a typical civil engineering environment, as she was in the past, when her job was likely to take her into uncomfortable, if not hazardous, situations. Of course, the increase in job safety and the decrease in hazards on the job also have been experienced by male members of the civil engineering profession as a result of the passage of comprehensive legislation designed to ensure the safety of employees in all occupations. Lastly, women are finding that many of the stereotypes that they themselves had pictured as members of the civil engineering profession simply do not exist. Any woman with an aptitude for science and mathematics, and an interest in planning, designing, and constructing physical facilities can find significant challenge and great rewards in civil engineering.

Today, an individual's performance is the most important basis for evaluation by colleagues and superiors in the field of civil engineering. Civil engineers are usually judged almost entirely on the basis of technical ability and communication skills.

EMPLOYERS' ATTITUDES

Another important consideration in evaluating any job offer is the attitude of the prospective employer toward the profession of civil engineering. Many employers, including consulting firms, governmental agencies, and large corporations, have demonstrated great support and respect for professional employees for many years. Such employers assist and support the new civil engineer in the development of her or his capabilities. Certainly, the development of professional attitudes among the technical staff in such organizations is the hallmark of a responsible employer.

In evaluating a potential job opportunity, the recent graduate or the civil engineer seeking a change of employment should determine whether or not the prospective employer will provide a satisfactory climate for professional development. Such a climate is created if the employer supports professional organizations and societies for engineers, such as the National Society of Professional Engineers and the American Society of Civil Engineers. The new graduate should attempt to determine if employees are encouraged to attend meetings and conventions of technical societies and professional civil engineering groups. He or she should ask if the expenses incurred by employees attending such meetings will be wholly or partially met by the prospective employer. He or she should also inquire if an employee can be an active member of committees or technical groups within a professional organization such as the American Society of Civil Engineers.

If the recent graduate obtains negative answers to the inquiries listed above and feels that employment with a particular company or agency will place her or him in an unprofessional atmosphere, he or she should consider employment elsewhere. A lack of a professional attitude toward engineering employees will probably be a negative factor in personal and professional development. This unpleasant situation would certainly outweigh other advantages such as high salary, excellent fringe benefits, or long vacations. The opportunities for professional fulfillment and advancement to more responsible positions should be much more important than financial rewards in one's career as a civil engineer.

EDUCATIONAL OPPORTUNITIES

The student in civil engineering who is completing a four-year or a five-year program may approach graduation with the feeling that her or his education is now complete. Nothing could be further from the truth. This feeling would be appropriate only for a person lacking in pride and ambition. The conscientious and ambitious civil engineer will attempt to learn something new every day on the job. However, with the ever-increasing complexity of the field of civil engineering, it is no longer possible to be well-informed even about a particular specialty such as structural engineering or geotechnical engineering merely through one's own efforts or through job experience. The new civil engineering graduate will find great difficulty even reading all the monthly journals and magazines associated with one particular area of civil engineering. For this reason, it is important for the new civil engineer, as well as the experienced civil engineer, to be active in constantly updating her or his education.

Continuing education can take several forms. The civil engineer may decide to participate in short courses or conferences which are held for several days in a particular location, usually on a university campus. On the other hand, he or she may choose to enroll in evening classes for an entire semester in order to develop deeper expertise in a particular subject. In some cases, it may be possible for the engineer to obtain permission from her or his employer to attend college classes during the day. These opportunities are important for the engineer to keep current her or his knowledge concerning specialized subjects. The opportunity to participate in these various forms of continuing education is an important asset in any job. When evaluating potential job opportunities, the new civil engineer should carefully investigate the opportunities in the offered position for participation in continuing education programs.

PERSONAL DEVELOPMENT

Professional development through continuing education and the gaining of experience in civil engineering activities will not be sufficient. Most civil engineers also will seek personal development through advancement to more responsible positions in their particular job assignments or in their company or organization. Obviously, advancement to positions of greater responsibility depends not only on the wishes and efforts of the individual but also on the judgment of the individual's immediate superiors and on the policies of the employing organization. In evaluating job opportunities, it is important for the civil engineer to carefully evaluate the opportunities for professional advancement. He or she should examine the organizational structure of a potential employer in order to determine whether or not it is possible to advance to more responsible positions after gaining experience and/or participating in further educational efforts, and if possible, he or she should try to estimate how rapidly he or she may advance to positions of more responsibility.

Greater responsibility in a job assignment can take a variety of forms. The young engineer may be given the chance to work with greater numbers of colleagues or with a larger labor force if he or she is engaged in the supervision of construction activities. On the other hand, working in an office environment he or she may be given the responsibility to plan and design more complex or important structures or systems. If he or she succeeds in demonstrating her or his capabilities, this personal advancement hopefully will be recognized by promotion to a higher grade or to a more responsible position with a corresponding increase in salary and other benefits.

In many large organizations such as federal governmental agencies, salary increases are given to engineers automatically with seniority at particular grade levels. Promotion to a higher grade level will come with increased experience on the job and with the demonstration of ability to handle more complex systems or to solve more difficult technical problems. Promotion to a higher grade level in a federal agency allows the individual to earn a new series of raises in salary and fringe benefits associated with that higher grade level, increasing seniority, and satisfactory performance of assigned duties.

In other organizations, such as consulting firms, increases in salary and benefits may not come automatically with seniority. However, the young civil engineer will likely have a greater opportunity in such firms to advance rapidly to senior technical or management positions. Nevertheless, in almost every organization, the civil engineer's advancement will be directly dependent upon the quality and quantity of the effort he or she expends on the job. In examining job offers, the new civil engineer should examine the potential for increased responsibility and higher salary which he or she is likely to receive for the investment of time and effort.

SALARIES AND FRINGE BENEFITS

The most obvious factor in evaluating a civil engineering job opportunity is the salary and fringe benefit package offered by the prospective employer. It is usually easy to compare one job offer with another on the basis of the salaries and fringe benefits associated with each position.

Salaries and fringe benefits are the most obvious attractions for any potential employment; however, in the long term, these factors are not as important as those discussed in the previous sections, such as personal interests and the possibilities for professional and personal development. On the other hand, salaries, fringe benefits, job security, and working conditions should be thoroughly considered. Starting salaries for civil engineering graduates today should not be a major factor in the evaluation of job offers, since the differences between starting salaries from one organization to another tend to be small and generally are balanced by differences in the fringe benefits offered by one employer as compared to those of another. Furthermore, differences in starting salaries usually disappear after several years employment with a given organization. In other words, two individuals hired at slightly different starting salaries generally tend to reach the same level of income in a specific organization if they have displayed equal capability and perseverance. The salary ranges which have been listed in previous chapters in this book indicate that, in any event, starting salaries for recent civil engineering graduates are likely to be higher than those of graduates in many other professional fields.

Fringe benefits differ from salary in that they are usually more important to employees with dependents, or with other circumstances which create special needs. A married engineer with a family may look more favorably on a job offer which includes payment of travel expenses incurred during the movement of family and belongings to the new place of employment, as compared to a job offer which does not include such a benefit. The health care and medical insurance provisions offered by a prospective employer are extremely important to married engineers with children, to persons with health problems, and to those who must support dependents with potential health problems. These factors are relatively unimportant, at least in the short term, to healthy, young, single college graduates. Obviously, the importance of health care and insurance benefits, as well as other benefits, will have varying degrees of importance to different individuals depending upon their particular circumstances. These benefits, however, must be evaluated carefully because they may have very great long-term significance.

Other benefits associated with employment as a civil engineer include unemployment insurance and retirement income. Almost all job assignments for civil engineers in the United States today will be covered by social security, which guarantees the engineer a modest retirement income. In almost all large engineering and technical organizations, supplementary retirement, profit sharing, or annuity plans are also available for all employees. The civil engineer considering a job offer should evaluate the benefits package available from a prospective employer, particularly with respect to the contribution provided by that employer to retirement plan payments.

JOB SECURITY

As mentioned earlier, there is good job security in civil engineering today. There has been, and should be, a continuing steady demand for civil engineers in today's technological society. Nevertheless, job security will vary somewhat depending upon the civil engineer's particular area of practice and upon the type of job secured. For example, the civil engineer employed by a federal government agency will be protected by the provisions of the federal civil service regulations. Although in some instances the civil engineer cannot gain employment in such an agency without passing a competitive examination, he or she is protected against arbitrary dismissal by this system. He or she can usually depend on subsequent promotions and raises in grade if performance is satisfactory and her or his progress agrees with the well-established procedures for promotion in the particular agency. A civil engineer will be discharged from a civil service position only if he or she is documented to be an unsatisfactory employee or is found guilty of illegal or immoral conduct. Consequently, employment in a federal agency, or in state and local agencies covered by civil service regulations, carries a great measure of job security, although some layoffs have occurred since 1981 when the jobs themselves were eliminated.

On the other hand, employment in private civil engineering firms is more vulnerable to economic fluctuations and the success of the particular private venture. Permanency of employment varies with the size and stability of the company or organization. In general, larger engineering organizations tend to be stable employers; however, civil engineers who are employed in organizations which are closely related to industrial production or to a particular element of the economy, such as defense spending, may face the possibility of temporary or permanent lay-offs.

Employment tends to be less stable in very small organizations. The civil engineer who is self-employed as a consulting engineer obviously has no guarantee of employment since he or she depends upon individual client–consultant agreements. Consultants experience a certain fluctuation in demand in that sometimes they are offered more work than they can possibly accomplish, while at other times they are idle for short periods, waiting to be contacted by clients. Against these dangers of possible unemployment, the self-employed consultant can count on one advantage: being one's own boss. He or she establishes her or his own salary and retirement program. This is an added responsibility and risk in self-employment. The civil engineer must weigh all of the possible advantages and disadvantages carefully before embarking on a career as a self-employed consultant or before becoming associated with a relatively small consulting firm.

PERSONAL INTERVIEW

The personal interview is a crucial step in the employment process for any engineer. For recent graduates, the personal interview may be the cause of considerable anxiety and concern. This interview need not be an occasion for anxiety; on the contrary, it should be an interesting and educational experience for both the applicant and the prospective employer.

The applicant should always remember that any interview is essentially a conversation between two people—the representative of the prospective employer and the applicant. If an engineer is being interviewed for a job in a governmental agency or a large corporation, it is quite likely that the interviewer will be a personnel professional whose specialty is the recruitment of technical employees. These representatives maintain continuous contact with the colleges and universities offering programs in civil engineering.

The interviewer may also be a civil engineer who is interested in personnel management, or he or she may be a personnel specialist without any extensive technical background. In any case, the applicant should always remember that the interviewer is the official representative of the prospective employer. Almost certainly, the interviewer will be well informed concerning the capabilities and needs of the organization he or she represents. Finally, the applicant should remember that the interviewer has been trained to evaluate prospective employees.

In most cases, the graduate is brought together with an interviewer by the college placement office. The time and place of the interview will be determined by the placement office in agreement with the representative of the employer. The placement office will usually publish notices concerning the times and places at which representatives of various employers will be present to conduct interviews. The prospective graduate can then register with the placement office for any interview in which he or she has an interest. Usually, the prospective employers will have furnished the placement office with literature concerning their organizations and their requirements; the applicant should obtain such literature and study it carefully. Then, he or she should make a list of questions which are not answered in the information furnished by the employer.

Interviewers attempt to maximize their efforts when they visit college campuses. Consequently their schedules are likely to be full and rather inflexible. For this reason, the applicant should be sure to arrive early for the interview. Tardiness for a job interview will certainly create a negative impression in the mind of the interviewer and may well lead to the loss of a desirable job offer. The applicant should be sure to dress appropriately; the interview season is a good time to buy a new suit. During the actual interview session, the applicant should allow the interviewer to guide and direct the flow of questions and the discussion. In almost all cases, the interviewer will have a list of routine questions asked of all job applicants. These questions will usually probe the engineer's interest in the work of the prospective employer, the educational courses or job experiences that were most rewarding, and the applicant's desires and ambitions in regard to future work. The applicant should try to respond to the interviewer's questions in a thoughtful and concise fashion, being careful not to wander into discussions of unrelated topics.

The interviewer usually will give the engineer an opportunity to ask questions; however, if the time for the interview is nearly gone and such an opportunity has not been given, the applicant should interrupt the discussion in order to ask any important questions he or she may have. In all cases, before the interview is at an end, the applicant should be sure that he or she completely understands the conditions of employment involved in a particular job situation. In some instances, a job may be offered at the conclusion of a single interview; in most cases, however, the first interview will lead only to a succession of further meetings. Additional meetings may lead to inspection trips. During these trips, the new civil engineer should be careful to remember that the treatment received during an inspection trip or plant visit may not be the same treatment he or she will receive as a regular employee. He or she should be careful to establish a good impression in the minds of prospective employers by prudent and careful conduct during inspection trips. It is important to remember that during such trips, not only is the applicant inspecting the prospective place of employment, but also he or she is being inspected by the prospective employers.

EVALUATING JOB OFFERS

After the new graduate has gone through a series of interviews and has received a number of job offers, he or she will be faced with the difficult decision of choosing the right job. In evaluating various job offers, he or she should review the factors listed in the first sections of this chapter. If possible, the applicant should discuss the offers with experienced friends or with colleagues at her or his place of business or with members of the faculty at school.

When the engineer finally has decided to accept a particular offer, he or she should advise the selected employer immediately and should inform the college placement office. At the same time, he or she should be sure to notify all other employers who have offered her or him positions that he or she has selected another offer. It is not necessary for the applicant to disclose which offer has been accepted unless he or she wishes to do so.

THE PATH TO ADVANCEMENT

After graduation, many civil engineers will obtain employment with a particular organization or agency and remain there throughout their entire professional careers. Alternatively, others will seek several changes in employment during their careers for personal or professional reasons. In either case, as an ambitious and conscientious person, the young civil engineer beginning a career should seek to advance professionally and personally. The path to advancement can have several different branches. For example, the civil engineer may move on to positions of greater responsibility in technical activities or may go into positions of responsibility in supervisory activities and management.

TECHNICAL SPECIALIZATION

In general, new civil engineering graduates initially will be employed to fulfill a basic technical position within an engineering firm or agency. The title they are given may be "junior engineer," "assistant engineer," or simply "engineer." If the graduate has an advanced degree, he or she may be employed at the next level up the ladder of professional advancement with a title such as "associate engineer." The successful engineer moves along a path of professional advancement to the point where he or she must select a career as a technical specialist, as a general practitioner, or as a manager. Usually this choice is faced at the level of "senior engineer" or "resident engineer." In planning and design offices, the "senior engineer" generally is responsible for coordinating the work of small numbers of junior and assistant engineers and auxiliary personnel such as technicians and draftsmen. The "resident engineer" usually is employed in construction to supervise the completion of a segment of a construction job rather than the entire job itself.

If the senior or resident engineer elects to advance professionally by becoming a technical specialist, he or she may be given a title such as "principal engineer." He or she may become a consultant to the general practitioners and management personnel within a given agency or organization. In general, the person who elects to become a technical specialist is the person who feels that he or she may not have the capabilities to be an effective manager or one who is unwilling to relinquish technical work for the increasing responsibilities in supervising and managing the work of others.

The general practitioner will follow a different path for professional development. He or she may rise from the level of senior design engineer or resident engineer to general engineer. In essence, advancement up the ladder of professional development requires supervising the work of an increasing number of subordinates. He or she must undertake more administrative work such as the preparation of reports on the activities of subordinates and must schedule individual workloads. Additionally, he or she may be involved in developing schedules and budgets for the work he or she supervises. In general, he or she is given a title such as "general project engineer" or "general design engineer" and rises to "principal engineer" or "principal project engineer." In consulting organizations, he or she may rise to a position as an associate of the firm, and then may assume more comprehensive administration duties as a vice-president or partner.

There are others who decide to forego direct technical activity in engineering for the challenge of management activities. Usually, this step is made at the level of senior or resident engineer. At this point, a civil engineer electing a career in management assumes more responsibility for coordinating and arranging the activities of a group of colleagues. He or she is engaged more frequently in making decisions on non-technical matters involving the allocation of personnel to specific tasks or in choosing objectives for the particular organization. He or she may assume successive titles such as "engineering superintendent," "branch chief," "engineering manager," or "division chief." If he or she reaches the executive level of management, he or she may have a title such as "director" or "president."

Advancement along any one of these various paths is dependent primarily on the individual's abilities, interest in the job, and perseverance in accomplishing her or his duties. In organizations where a civil service system governs promotions and advancement, employees will be promoted on the basis of their technical expertise and achievements as well as their accumulated seniority. Their ability to accept more complex and demanding tasks and to carry out those tasks successfully also will be taken into account. In some organizations, the employee must take examinations to qualify for promotion. In other firms or agencies, a board of experienced senior engineers will review an employee's credentials and work experience and recommend promotions.

In organizations without civil service procedures, generally a somewhat parallel or similar administrative procedure has been established to show the employee what must be done to achieve promotion and advancement. Obviously, in small firms, and especially in consulting organizations, promotion and advancement may be dependent upon the opinion of one or two senior individuals such as the president or owner of the consulting firm. In any case, the most important requirements for advancement in the profession of civil engineering are a demonstration of technical competence and willingness to spend the required time and effort necessary to accomplish assigned tasks and responsibilities.

In a civil engineering career, there is no substitute for hard work. However, an individual may find great encouragement through the establishment of close contacts with fellow engineers, not only in her or his own place of employment but in other agencies and organizations through membership in professional organizations and societies. Memberships in such organizations can be very effective in broadening the professional development of the young engineer and presenting the experienced engineer with more opportunities for the exercise of her or his talents and capabilities. Such organizations are described in Chapter 6.

PROFESSIONAL ORGANIZATIONS IN CIVIL ENGINEERING

Various organizations exist through which the reader can discover more about careers in engineering. Nearly 200 engineering or related societies are listed in the Directory of the Engineers Joint Council. The services of many of these organizations are available to both high school and college students, as well as to practicing engineers. Some of these organizations are open for membership to all engineers without regard to their area of professional specialization, while others, such as the American Society of Civil Engineers, are for members who practice a particular type of engineering. During college, a student can join a number of honorary, social, and professional organizations affiliated and involved with civil engineering. A graduate engineer can become registered to formally recognize his or her capabilities as a practicing professional engineer, and he or she can become a member of one or more of the professional organizations specifically formed for civil engineers.

ORGANIZATIONS IN HIGH SCHOOL

Many American engineering societies exist for the basic purpose of disseminating information. Among the numerous types of information available from these societies, including the American

129

Society of Civil Engineers, is guidance material for high school students considering a career in engineering. High school students can obtain this information concerning engineering careers from their counselors or teachers or from local representatives or members of the professional societies. The National Society of Professional Engineers is deeply involved in career orientation programs specifically directed toward high school students. Representatives of this society regularly conduct programs throughout the country. In such programs, high school students visit engineering, industrial, and technical organizations to see engineers and technologists at work. These students are given an opportunity to investigate the work of the engineers as well as to meet and talk with practicing engineers. Interested students or counselors should contact the local chapter of the National Society of Professional Engineers for further information about such programs in their localities.

In addition to the orientation programs conducted by the National Society of Professional Engineers, a separate program, the Junior Engineering Technical Society (JETS), also exists for the purpose of furnishing information to high school students who are interested in careers in engineering. This organization was founded in 1950 to sponsor extracurricular programs for high school students in both rural and urban areas. One of its primary activities is the sponsorship of JETS chapters in junior and senior high schools. These chapters, in turn, sponsor extracurricular clubs under the supervision of a faculty advisor assisted by a volunteer professional engineer from the community.

The typical JETS chapter or club holds regular meetings at which members explore various aspects of engineering. The club members may visit local industries, consulting firms, and government agencies to discuss their interests with engineers and scientists in practice. Technical programs, projects, and other group activities are part of most clubs' schedules. Information concerning membership can be obtained from any local chapter of that organization or from the JETS National Office, 345 East 47th Street, New York, New York 10017. JETS chapters have been formed in many cities and towns in the United States and have proven to be a very worthwhile means to acquaint high school students with the career opportunities in civil engineering.

In addition to the high school clubs and programs sponsored by engineering societies, there are many other science and mathematics clubs that the prospective civil engineering student will find to be educational as well as enjoyable. These clubs frequently have programs and speakers that are valuable sources of information for young people who are trying to choose a career.

COLLEGE PROFESSIONAL AND SOCIAL ORGANIZATIONS

Students in a civil engineering curriculum at a university or college can become members of several different organizations associated with their particular field of study. Almost all of the national engineering societies, including the American Society of Civil Engineers, have student chapters in colleges and universities throughout the United States. The student chapter of ASCE on the campus will be sponsored by a local section of the national organization. In addition to the very interesting programs and field trips that the student chapter officers arrange for chapter members, the students frequently will be invited to participate in the activities of the local section. These activities will include programs with prominent engineering speakers and tours behind the scenes of outstanding civil engineering projects in the area.

The National Society of Professional Engineers (NSPE) also sponsors student chapters in many engineering schools. These clubs are open to students from all branches of engineering and have programs of interest to all engineering undergraduates.

The civil engineering student may be invited to become a member of Chi Epsilon, an undergraduate honor society for outstanding civil engineering students. Chapters of Chi Epsilon are found at many universities. The civil engineering student may also be invited, as an outstanding scholar, to become a member of Tau Beta Pi, the national honorary engineering society. Election to Tau Beta Pi, for an engineer, is equivalent to election to Phi Beta Kappa for a scholar in the arts and sciences. Members, both male and female, are chosen for this society on the basis of outstanding potential as a member of the profession of engineering, as well as on their performance as undergraduate students.

In addition to these honorary and professional organizations, there are many social and service organizations on every campus which are open to civil engineering students. The prospective engineer would be wise to consider joining one or more of these organizations in order to gain social contact with students from diverse backgrounds, interests, and fields of study. Among these groups are the Greek fraternities and sororities, religious clubs, military veteran groups, political clubs, and the many types of musical associations open to student members.

For members of minority groups and for women, special organizations have been formed to fulfill identified needs. Women interested in careers in engineering or who already are practicing in the field can become members of the Society of Women Engineers. Information on this society is available to interested women from the Society at the United Engineering Center, 345 East 47th Street, New York, New York 10017. In addition to this organization for women, a number of other organizations have been formed to respond to the needs of various ethnic and racial groups whose members practice in the field of engineering.

PROFESSIONAL LICENSING AND REGISTRATION

Of all engineering professionals, the civil engineer deals more directly with, and is more responsible for, maintaining public health and safety than any other type of engineer. For this reason, it is important that there be laws requiring the examination of the qualifications of civil engineers before they are permitted to practice in capacities where public safety is their concern. All fifty states and the District of Columbia currently have laws requiring the licensing or registration of professional engineers, including civil engineers. These laws require that members of registration boards, established for this purpose by the various state legislatures, secure evidence of education and experience for each applicant and determine whether or not the applicant fully understands the basic principles of science and technology necessary for the safe practice of engineering in that particular state. While the requirements in each state differ in accordance with the specific law of each state, registration is normally granted only after completion of four years acceptable engineering experience and passing a one- or two-day comprehensive examination.

The primary justification for registration and licensing of engineers is the protection of the public. The requirement for licensing was well stated some years ago by the judge of a Utah court in the case of *Clayton v. Bennett* (298 P. 2d 531) as follows:

> It has been recognized since time immemorial that there are some professions and occupations which require special skill, learning, and experience with respect to which the public ordinarily does not have sufficient knowledge to determine the qualifications of the practitioner. The layman should be able to request such services with some degree of assurance that those holding themselves out to perform them are qualified to do so. For the purpose of protecting the health, safety, and welfare of its citizens, it is within the police power of the State to establish reasonable standards to be complied with as a prerequisite to engaging in such pursuits.

In this regard, it is recommended that all civil engineering graduates become registered as soon as possible after graduation. There are various reasons for this recommendation; however, the two most compelling reasons are: 1) courts of law generally do not recognize a person as an engineer unless that person is registered (this can be a serious obstacle should the person ever wish to testify as an expert witness); and, 2) while the activity that an engineer is currently engaged in may not require registration, it is possible that at some point in her or his career he or she may wish to assume a position in which registration is required.

As pointed out previously, each state has its own specific registration law. However, the most common requirement for registration at the present time is the successful completion of two one-day comprehensive examinations. The first day's examination usually covers the basic fundamentals of engineering as taught in an ABET-accredited curriculum. Many states permit this first examination to be taken just prior to, or immediately after, graduation. When a candidate successfully passes this examination, he or she is granted a certificate designating her or him as an Engineer-In-Training (EIT). The second examination may be taken only after the applicant fulfills the particular state requirement for minimum number of years of practical experience, which is usually four years. Upon completion of this second requirement (the years of experience plus the second examination), the graduate engineer is registered and granted a license to practice engineering in that state.

Once registration is granted, further examination is not required. However, several state registration boards have considered the adoption of a requirement that periodically an engineer may be required to show proof of some degree of continuing engineering education before renewal of the engineer's license. These requirements, if adopted, will probably be similar to those now in existence for the medical, law, and teaching professions.

MEMBERSHIP IN PROFESSIONAL SOCIETIES

As previously stated, there are more than 200 engineering societies or related groups in the United States today. The graduate civil engineer will find that professionally it will be profitable to join several of these societies, including at least one devoted to the field of civil engineering. The primary purpose of most professional organizations is the exchange of information among its members to their mutual benefit.

The American Society of Civil Engineers (ASCE) is the oldest engineering society in the United States. Among its objectives, as stated by the society itself, are:

1. To encourage and publicize discoveries and new techniques throughout the profession.
2. To afford professional associations and develop professional consciousness among civil engineering students.
3. To further research, design, and construction procedures in specialized fields of civil engineering.

4. To give special attention to the professional and economic aspects of the practice of engineering.
5. To enhance the standing of engineers.
6. To maintain and improve standards of engineering education.
7. To bring engineers together for the exchange of information and ideas.

The National Society of Professional Engineers (NSPE) has as its objective: "The promotion of the profession of engineering as a social and as an economic influence vital to the affairs of men and women of the United States." It is one of the largest societies in this country with membership open to all registered engineers. NSPE is headquartered in Washington, D.C., and takes a very active interest in any national legislation that affects engineers or the profession of engineering.

There are many other societies that the practicing civil engineer may also wish to join in order to associate with other engineers and professionals doing similar kinds of work. Among these latter types of societies would be such groups as the American Public Works Association, the American Water Works Association, the American Association of State Highway and Transportation Officials, the American Concrete Institute, the Transportation Research Board, the Water Pollution Control Federation, etc.

Membership requirements will vary from society to society; however, many groups such as the ASCE have student membership grades from which the student may transfer to associate membership upon graduation. The primary reason for belonging to a society of any type is to participate in its activities, and to contribute to the exchange of professional information so vital to any profession's existence. Each prospective civil engineer is encouraged to investigate the student society memberships available while completing undergraduate studies.

Mass transportation systems, such as the one shown here, are designed by civil engineers. (American Society of Civil Engineers photo)

A FINAL WORD

In the preceding chapters, the profession of civil engineering and all the wide range of activities and specialties within that profession have been described. It was not possible to include all of the job assignments open to civil engineers, but many typical engineering activities have been described. Anyone interested in a career in civil engineering should have discovered from those descriptions that an opportunity exists today to work in a variety of technical specialty areas within civil engineering, for a more-than-adequate salary and under very favorable working conditions.

The preparation involved in a college curriculum in civil engineering and, for that matter, a degree in civil engineering do not guarantee subsequent advancement in the profession. However, civil engineers who are ambitious and persevere in their efforts almost surely will advance professionally and personally. The civil engineer's responsibilities and rewards will increase with experience on the job. Additionally, the ambitious civil engineer will find many opportunities to move into related fields of engineering or to advance into management positions supervising both technical and non-technical personnel. A basic education and experience in civil engineering can also provide a foundation for a rewarding career in engineering sales. On the whole, a career in civil engineering offers great promise; however, it requires dedication and hard work and possesses certain limitations.

For example, civil engineers deal principally with structures and devices rather than with people. An engineer analyzes and designs systems consisting of natural forces and material goods and does not

137

always deal principally with individuals or communities. Obviously, however, the civil engineer's activities have great impact on the welfare of both individuals and communities. Nevertheless, the civil engineer relies on background and education in the physical and natural sciences, particularly mathematics, to solve technical problems in the real world. He or she is not necessarily interested in searching for absolute truths as is the scientist, nor does he or she deal with specific mechanics of problem solving as does the technician. Rather, he or she attempts to analyze general situations; to develop and design systems to satisfy needs and solve problems; and to oversee and supervise the implementation of these designs.

In general, civil engineers are well-paid professionals. Engineering careers offer quite adequate job security with good salaries and fringe benefits for those who are capable. In comparison to other engineering specialities, average starting salaries for civil engineers are lower. However, employment opportunities in civil engineering are not subject to major fluctuations with economic conditions as are other branches of engineering. Additionally, a significant fraction of the total number of civil engineers find employment as consultants; when the civil engineer is employed in this fashion, the potential for attaining salaries much higher than the average are greatly enhanced. In any case, the civil engineer will usually be able to see the designs and solutions he or she has prepared implemented in concrete and steel or in a similar material fashion. This direct evidence of productive effort is one of the most rewarding features of a civil engineering career. If, on the other hand, you derive more satisfaction from dealing with the intangible personal problems of your fellow human beings, you may feel frustrated in working as an engineer where you must concentrate on material things rather than on people and personal relationships.

If you are considering a career in civil engineering, examine very carefully the information presented in the first and third chapters of this book which describe the functions of civil engineering and the technical specialties within the general field of civil engineering. Review the job duties, salaries, and working conditions, and the general potential for advancement which are described in Chapters 4 and 5. If the activities described in this book seem attractive and appealing to you, and if you have a strong capability in mathematics and the physical sciences, it is likely that you will find great satisfaction in a career as a civil engineer. If you feel that you have an inclination toward this field of endeavor, carefully review the material in Chapter 2 describing the education necessary for practice as a civil engineer. Be sure to establish the correct foundation for a college education in civil engineering by enrolling in the proper high school courses in mathematics and physical science. Carefully choose an accredited engineering school and explore all possible opportunities for financial assistance at that school. Extend your explorations to several schools if you choose. After you have made your choice and have entered a university or college to pursue an education in civil engineering, carefully review your progress to date and be sure that you have made the correct career choice. If you remain satisfied and you successfully complete your college education, we look forward to having you join us as colleagues in the great profession of civil engineering.

APPENDIX A

GENERAL TECHNICAL PUBLICATIONS

APWA Reporter
American Public Works Association
1313 East 60th Street
Chicago, IL 60637

Civil Engineering
American Society of Civil Engineers
345 East 47th Street
New York, NY 10017

Certified Engineering Technician
American Society of
  Certified Engineering
  Technicians
2029 K Street, N.W.
Washington, D.C. 20006

Engineering News-Record
McGraw-Hill Book Company, Inc.
330 W. 42nd Street
New York, NY 10036

Consulting Engineer
Consulting Engineers Council
  of the U.S.A.
1155 15th Street, N.W.
Washington, D.C. 20005

APPENDIX B

RECOMMENDED READING

American Society of Civil Engineers, *The Civil Engineer: His Origins,* ASCE, New York, NY, 1970.

U.S. Department of Labor, *Occupational Outlook Handbook,* Superintendent of Documents, Washington, D.C., 1986.

Finch, J. K., *The Story of Engineering,* Doubleday and Company, New York, NY, 1960.

Florman, S.C., *The Existential Pleasures of Engineering,* St. Martin's Press, New York, NY, 1976.

Furnas, C. C. and J. McCarthy, *The Engineer,* Life Science Library, Time-Life Books, Chicago, IL, 1966.

Greenwald, S. M. and I. R. Wecker, "Transfers to Engineering, $2 + 2 = 6$?", *Engineering Education,* Vol. 65, No. 8 (May, 1975).

Hauck, G. F., "Technology/Engineering Articulation," *Engineering Education,* Vol. 62, No. 5 (February, 1972).

Jacobs, D. and A. E. Neville, *Bridges, Canals and Tunnels,* The American Heritage Publishing Company, New York, NY, 1968.

Kirby, R. S., S. Withington, A. Darling, and F. G. Kilgour, *Engineering in History,* McGraw-Hill Book Company, New York, NY, 1956.

Minorities in Engineering: a Blueprint for Action, The Planning Commission for Expanding Minority Opportunities in Engineering, Alfred P. Sloan Foundation, 1971.

Morgan, A. E., *The Making of the TVA,* Prometheus Books, Buffalo, NY, 1974.

Pannell, J. P. M., *Man the Builder,* Crescent Books, New York, NY, 1964.

Peden, I. C., "Women in Engineering," *The World of Engineering,* Ed., J. R. Whinnery, McGraw-Hill Book Company, New York, NY, 1965.

Penzias, W. and M. W. Goodman, *Man Beneath the Sea,* Wiley-Interscience, New York, NY, 1973.

Sandstrom, G. E., *Man the Builder,* McGraw-Hill Book Company, Inc., New York, NY, 1970.

Smith, R. J., *Engineering as a Career,* McGraw-Hill Book Company, Inc., New York, NY, 1969.

Sullivan, G., *How Do They Build It?,* Westminster Press, Philadelphia, PA, 1972.

Walker, E., "Engineers and/or Scientists," *Engineering Education,* Vol. 61, No. 5 (February, 1961).

ADDITIONAL INFORMATION

A very complete and useful set of brochures and short articles on various aspects of engineering careers is available from JETS, Inc., a non-profit youth activity organization. A partial listing is given below.

Individual requests for single copies, from students and teachers, for guidance materials marked with an asterisk will be honored at no charge. All requests for these materials should be made to:

JETS-Guidance
345 East 47th Street
New York, NY 10017

*EC-13 After High School—What? 9pp.

*EC-15 A Primer on Financial Aid for Students. 1979.

*EC-62 Engineering: Creating a Better World. 1970. 20 pp.

EC-63 New Careers in Engineering Technology. 1975. 20 pp.

EC-66 Engineering. 1980. 36 pp.

EC-69 The Engineering Team. 1974. 16 pp.

EC-70 Womengineer. 1974. 20 pp.

EC-74 Engineering. 1983. 40 pp.

EC-79 Building an Engineering or Technical Career. 1978.

EC-79A Transparencies for Building an Engineering or Technical Career.

*EC-91 **Engineering Team Members.** 1976.

*EC-92 **Women in Engineering.** 1978.

*EC-93 **Engineering Education.** 1978.

*EC-94 **Minorities in Engineering.** 1979.

*EC-95 **Women in Technology.** 1979.

ABET PUBLICATIONS

(Available through arrangement with the Association Board for Engineering and Technology)

AB-20 Accredited Programs Leading to Degrees in Engineering. 1984.

AB-21 Accredited Programs Leading to Degrees in Engineering Technology. 1984.

EC-80 THE JETS PROGRAM. 8 pp.

EC-82 JETS REPORT. 9 issues, 8-pp. newsletter.

EC-84 JETS CHAPTER HANDBOOK. 1983. 80 pp.

EC-85 JETS PROGRAM AIDS. 1985. 64 pp.

EC-86 JETS STARTER KIT.

ACCREDITED SCHOOLS

More than 190 universities and colleges currently offer programs in civil engineering in the United States.

The prospective student in civil engineering will do well to investigate in depth any school that he or she is considering attending. As recommended in Chapter 2, such an investigation should include an evaluation of the professional background of the faculty members, an investigation of the school facilities and laboratory equipment available to the engineering students, and as much additional information about the school as the student can gather. This type of investigation is somewhat difficult for the high school student who is not familiar with handling information of this sort or evaluating engineering programs. In order to assist the public in evaluating engineering schools, professional organizations in engineering have gathered together to form an agency to accredit engineering schools. These professional agencies look closely at the program of course work offered at each school. The combined accreditation group is known as the Accreditation Board for Engineering and Technology.

The Accreditation Board for Engineering and Technology does not rank schools or their programs; rather, they publish a listing of those programs which are acceptable to all of the members of the technical societies which banded together to form the Accreditation Board. Groups of examiners travel to the individual institutions to conduct accreditation inspections. The accreditations are based on a review of course offerings, an evaluation of faculty capabilities, and on-site inspections of the physical facilities at each institution. A student considering a career in civil engineering can be confident of receiving an adequate preparation in civil engineering at any school accredited by ABET. A complete listing of accredited programs may be found in ABET report AB-20 entitled *Accredited Programs Leading to Degrees in Engineering.*

VGM CAREER BOOKS

OPPORTUNITIES IN
Available in both paperback and hardbound editions
Accounting Careers
Acting Careers
Advertising Careers
Airline Careers
Animal and Pet Care
Appraising Valuation Science
Architecture
Automotive Service
Banking
Beauty Culture
Biological Sciences
Book Publishing Careers
Broadcasting Careers
Building Construction Trades
Business Communication Careers
Business Management
Cable Television
Carpentry Careers
Chemical Engineering
Chemistry Careers
Child Care Careers
Chiropractic Health Care
Civil Engineering Careers
Commercial Art and Graphic Design
Computer Aided Design and Computer Aided Mfg.
Computer Science Careers
Counseling & Development
Dance
Data Processing Careers
Dental Care
Drafting Careers
Electrical Trades
Electronic and Electrical Engineering
Energy Careers
Engineering Technology Careers
Environmental Careers
Fashion Careers
Federal Government Careers
Film Careers
Financial Careers
Fire Protection Services
Fitness Careers
Food Services
Foreign Language Careers
Forestry Careers
Gerontology Careers
Government Service
Graphic Communications
Health and Medical Careers

High Tech Careers
Hospital Administration
Hotel & Motel Management
Industrial Design
Insurance Careers
Interior Design
Journalism Careers
Landscape Architecture
Law Careers
Law Enforcement and Criminal Justice
Library and Information Science
Machine Trades
Magazine Publishing Careers
Management
Marine & Maritime
Materials Science
Mechanical Engineering
Microelectronics
Modeling Careers
Music Careers
Nursing Careers
Nutrition Careers
Occupational Therapy
Office Occupations
Opticianry
Optometry
Packaging Science
Paralegal Careers
Paramedical Careers
Part-time & Summer Jobs
Personnel Management
Pharmacy Careers
Photography
Physical Therapy Careers
Podiatric Medicine
Printing Careers
Psychiatry
Psychology
Public Relations Careers
Real Estate
Recreation and Leisure
Refrigeration and Air Conditioning
Religious Service
Robotics Careers
Sales & Marketing
Secretarial Careers
Securities Industry
Sports & Athletics
Sports Medicine
State and Local Government
Teaching Careers
Technical Communications
Telecommunications

Theatrical Design & Production
Transportation
Travel Careers
Veterinary Medicine Careers
Vocational and Technical Careers
Word Processing
Writing Careers
Your Own Service Business

WOMEN IN
Communications
Engineering
Finance
Government
Management
Science
Their Own Business

CAREERS IN
Accounting
Business
Communications
Computers
Health Care
Science

CAREER DIRECTORY
Occupational Outlook Handbook

CAREER PLANNING
How to Get People to Do Things Your Way
How to Have a Winning Job Interview
How to Land a Better Job
How to Write a Winning Résumé
Life Plan
Planning Your Career Change
Planning Your Career of Tomorrow
Planning Your College Education
Planning Your Military Career
Planning Your Own Home Business
Planning Your Young Child's Education

SURVIVAL GUIDES
High School Survival Guide
College Survival Guide

 VGM Career Horizons
A Division of National Textbook Company
4255 West Touhy Avenue
Lincolnwood, Illinois 60646-1975 U.S.A.